Behavioral Ecology of the Eastern Red-backed Salamander

Behavioral Ecology of the Eastern Red-backed Salamander

50 Years of Research

BY ROBERT G. JAEGER,

BIRGIT GOLLMANN,

CARL D. ANTHONY,

CAITLIN R. GABOR

AND

NANCY R. KOHN

Oxford University Press is a department of the University of Oxford. It furthers the University's objective of excellence in research, scholarship, and education by publishing worldwide. Oxford is a registered trade mark of Oxford University Press in the UK and certain other countries.

Published in the United States of America by Oxford University Press
198 Madison Avenue, New York, NY 10016, United States of America.

Library of Congress Cataloging-in-Publication Data
Names: Jaeger, Robert, author.
Title: Behavioral Ecology of the Eastern Red-backed Salamander / by Robert G. Jaeger, Birgit Gollmann, Carl D. Anthony, Caitlin R. Gabor, and Nancy R. Kohn.
Description: New York, NY : Oxford University Press, 2016. |
Includes bibliographical references and index.
Identifiers: LCCN 2015050051 | ISBN 9780190605506
Subjects: LCSH: Plethodon cinereus—Behavior. | Plethodon cinereus—Ecology.
Classification: LCC QL668.C274 J34 2016 | DDC 597.8/5—dc23
LC record available at https://lccn.loc.gov/2015050051

For Professor Jane Brockmann, who initially suggested and encouraged this book, and Professor Murray Itzkowitz, whose juxtaposition of field and laboratory experimentation was an inspiration for our research

CONTENTS

LIST OF FIGURES

ACKNOWLEDGMENTS

We thank Günter Gollmann (Universität Wien) for useful discussions while we climbed Mount Schneeberg in the high Austrian Alps with the help of the "Salamander" train. We also thank Terry Kohn for her assistance in typing large portions of this book and Jenny Thibodeaux for computer assistance. Lauren Mathews provided advice for section 10.4 and Andrea Aspbury, Douglas Fraser, Günter Gollmann, and Sharon Wise provided critical advice as we composed the manuscript. Nancy Kohn, Caitlin Gabor, Carl Anthony, and Bob Jaeger supplied the photographs. The research reviewed here was supported by the following universities and research stations: University of Maryland at College Park and Shenandoah National Park (SNP; 1965–1971); University of Wisconsin at Madison and SNP (1971–1974); State University of New York at Albany and SNP (1974–1980); University of Louisiana at Lafayette (1981–2008); and Mountain Lake Biological Station of the University of Virginia (1981–2015). Our research was shaped by three philosophers, K. Popper, J. Platt, and I. Lakatos, whose philosophies of science led the path to the progressive research program described herein with the ever-fascinating eastern red-backed salamander, *Plethodon cinereus*.

ABOUT THE AUTHORS

Robert G. Jaeger, Department of Biology, Utica College, Utica, New York, USA; 100 Simon Latour Road, Carencro, Louisiana, USA

Birgit Gollmann, Fakultät für Lebenswissenschaften, Universität Wien, Wien, Austria

Carl D. Anthony, Department of Biology, John Carroll University, University Heights, Ohio, USA

Caitlin R. Gabor (corresponding author), Department of Biology, Texas State University, San Marcos, Texas, USA, gabor@txstate.edu

Nancy R. Kohn, Department of Biology, The College of New Jersey, Ewing, New Jersey, USA; Department of Biology, University of Missouri-St. Louis, St. Louis, Missouri, USA

Behavioral Ecology of the Eastern Red-backed Salamander

1

Prelude

1.1 BOB JAEGER MEETS THE EASTERN RED-BACKED SALAMANDER, *PLETHODON CINEREUS*

I take the liberty here to write in the first-person singular to convey how the following research program began. During the late 1950s, I was an undergraduate student at the University of Maryland at College Park, majoring in zoology. My initial intention was to train in ichthyology, until I realized that I was deathly fearful of entering waters deeper than my ankles! Not many species of fishes (especially sharks) swim in waters only 9 cm deep, so my thoughts quickly changed to amphibians, many of which live in shallow water. Frogs and toads had fascinated me during childhood while wandering in the forests of rural Maryland, so I sought a faculty member who studied anurans (frogs). Luckily, during my junior and senior years, Dr. Richard Highton hired me as a research assistant. Dick Highton is an authority of salamandrine taxonomy, systematics, and geographic variation with a particular interest in the family Plethodontidae, which includes *Plethodon cinereus*, the eastern red-backed salamander (hereafter the red-backed salamander). Salamanders are not frogs, but they were close enough for me back then. Dick Highton frequently led his flock of graduate and undergraduate students on collecting trips throughout Maryland, Delaware, Pennsylvania, West Virginia, and Virginia. He seemed to have a special passion for red-backed salamanders, which can be seen by his approximately 100,000 preserved specimens that now reside in the US National Museum.

I was then not interested in taxonomy, even less so after soaking my hands, for two years, in formaldehyde while preserving Dick Highton's salamanders. However, I was curious about the thousands of red-backed salamanders that we found (and then collected) in the forests. Three observations struck me as odd.

First, many forests had far greater densities of this species than of other co-occurring species of caudate (tailed) amphibians. (Later, Burton & Likens, 1975, found that in a forest in New Hampshire, the red-backed salamander comprised 93.5% of the total biomass of all species of salamanders, with 885 to 2,367 red-backed salamander per hectare.) I was puzzled by how one species of salamander could maintain such large densities.

Second, while on a "good collecting day" I would find a red-backed salamander under nearly every rock in a forest, even large rocks usually housed only one salamander, except during the spring and autumn courtship seasons when pairs were frequently found together under rocks. This too seemed odd, because large rocks clearly covered enough soil to house more than one or two individuals.

Third, in some forests, the geographic distribution of the red-backed salamander abruptly ended and the distribution of another same-size congener began. Why would two species so confront each other with very little area of species' overlap? This was how I first met the red-backed salamander: puzzled by these three observations.

As an undergraduate student, I was woefully ignorant of ecological theories and empirical studies, because the course program then largely focused on taxonomy, anatomy, genetics, embryology, and physiology. So I progressed on to the University of California at Berkeley in the early 1960s for a master's degree. At Berkeley, when not dodging or participating in demonstrations, I was able to enjoy several ecology courses that expounded on the then-current theories of (1) population regulation, (2) intraspecific competition, and (3) interspecific competition. Also, I was fortunate enough to take Peter Marler's course in (4) animal behavior. These four topics seemed to be pertinent, somehow, to my previous observations of red-backed salamanders and their densities, intraspecific distributions, and interspecific interactions.

Therefore, with master's degree in hand, I returned to College Park, Maryland, for a doctoral degree where I worked with Dick Highton as an advisor, because he could tell me where the salamanders were and because the department had recently hired an ecologist and behaviorists with whom I filled out my doctoral committee. The resulting PhD ended with four published papers concerning interspecific competition between the red-backed salamander and the Shenandoah salamander, which are summarized in section 2.1. However, this research seemed to me to be incomplete, because the interspecific ecological interactions that I had documented seemed to have a behavioral underpinning, which I was still too naïve to understand.

This lack of behavioral knowledge sent me to the University of Wisconsin at Madison in 1971, blissfully supported in postdoctoral research by a three-year grant from the National Science Foundation to Dr. Jack P. Hailman. Jack was a superb ethologist who taught the basics of experimental designs within a behavioral context to me. The postdoctoral research concentrated on phototactic behavior of anurans, so at last I had found a niche in the biology of frogs and toads! Yet I was also able to continue studies of the red-backed salamander during summers, back in Virginia at my doctoral research site (Fig. 1.1).

Finally, in 1974, armed with a total of 12 publications, it was time to earn an honest living with a viable salary, so I obtained a position at the State University of New York at Albany, later moving to the University of Louisiana at Lafayette in 1981. These two universities required little teaching (one undergraduate and one graduate seminar course per year), so much of my time could be devoted to research with the red-backed salamander. Most of the studies summarized in section 2.2 through chapter 10 were conducted at these two

Figure 1.1 A young Bob Jaeger, c. 1973, in Shenandoah National Park, Virginia.

institutions (many behavioral experiments) and at research sites in northern New York and in Virginia (many ecological experiments). A large number of my doctoral, master's, and undergraduate students chose to test (often seemingly wild) hypotheses concerning the behavioral ecology of the red-backed salamander, which led to many of the 114 publications plus three unpublished theses and six partially unpublished dissertations discussed in this book.

Once I reached retirement in 2006, I assumed that my remaining years could be devoted to my other pleasures: trips to operas and ballets in the United States and Europe. While this has come to pass, research with colleagues and former students continues to generate new manuscripts about red-backed salamanders. Then, in 2012, Jane Brockmann encouraged me to compose this book, integrating 50 years of research with the red-backed salamander into a long story of discoveries by my research group. Jane has been a dear friend since our graduate student days in the late 1960s at the University of Maryland, so I was quite flattered by her suggestion, but not excited about writing a book alone. The venture was saved when Nancy Kohn, Birgit Gollmann, Caitlin Gabor, and Carl Anthony offered to write some other sections of the book. The five of us hope that this book will encourage other behavioral ecologists to discover more surprises still concealed by the mysterious red-backed salamander and, perhaps, by other species of tailed amphibians.

1.2 AN INTRODUCTION TO RED-BACKED SALAMANDERS

The natural history, ecology, and reproductive behavior of red-backed salamanders have been well summarized by Petranka (1998), Wells (2007), and Anthony and Pfingsten (2013). Therefore, here we note only a few characteristics of the species that were pertinent to our research program.

The red-backed salamander (Fig. 1.2) is a small salamander about the thickness of a pencil. Its maximum length varies geographically, but at our research sites, fully grown adults reach approximately 45 mm snout-to-vent length and 90 mm total length (including its long tail). Red-backed salamanders are lungless, as are all species in the family Plethodontidae, so respiration occurs exclusively through skin surfaces. Consequently, plethodontid salamanders have permeable skin requiring a moist environment, as do nearly all amphibians, to prevent desiccation. It is this requirement that largely shapes the ecology of the genus *Plethodon*, because these species are strictly terrestrial. All life stages of the red-backed salamander are found in the leaf litter of the forest floor (egg, neonate, juvenile, and adult), where individuals forage for invertebrates, but they frequently are found under rocks and logs when the leaf litter dries.

Figure 1.2 An adult, female red-backed salamander, *Plethodon cinereus*, uncovered in a deciduous forest at Mountain Lake Biological Station, Virginia. She is ~45 mm snout-to-vent length and ~90 mm total length, lying near a damp log, and in the resting and alert (head raised) posture (front of trunk raised). Her large eyes provide enhanced vision under dim light in the leaf litter. Note that her tail is thick (indicating stored fat from successful recent foraging) and intact (not autotomized). Her red stripe extends down only 40% of the tail, which probably means that the terminal 60% had been autotomized (1+ years previously) and then regrown. Tail autotomy can occur during a predatory attack, in which the tail, still thrashing, distracts the predator's attention, or during a biting attack from a territorial rival, during which the attacker usually eats the autotomized tail. This is a striped morph of *P. cinereus*, but in this population individuals vary from full-length stripe to spotty blotches of red, orange, yellow, or tan to no stripe at all (the leadback morph).

During extremely dry and freezing winter conditions, they move underground following crevices in the soil; this species is not adapted to dig its own burrows.

The geographic distribution of red-backed salamanders is enormous compared to all other species of *Plethodon*, except for the much larger *P. glutinosus* (northern slimy salamander; see Petranka, 1998, for range maps and taxonomic controversies concerning *P. glutinosus*). Red-backed salamanders are found in most, but not all, cool, closed canopy forests throughout northeastern North America, from southeastern Canada southward to northern North Carolina and from the Atlantic Coast northwestward to the northern Mississippi River. Due to the relatively low efficiency of gas exchange by cutaneous respiration, these salamanders require cool conditions, and thus southern populations are constrained to mountainous areas.

Several other species of *Plethodon* occur within the geographic range of red-backed salamanders. Some of these species have only tiny distributions (e.g., *P. aureolus, P. hubrichti, P. nettingi, P. punctatus*, and *P. shenandoah*), while other species are more widely distributed (e.g., *P. dorsalis, P. glutinosus, P. hoffmani, P. richmondi, P. wehrlei*, and *P. welleri*), according to range maps in Petranka (1998). We were particularly interested in areas where red-backed salamanders overlap only slightly with other species of a similar size, and in chapter 6 we summarize our research concerning competition between *P. cinereus* and *P. hubrichti, P. hoffmani*, and *P. shenandoah*.

However, red-backed salamanders often coexist in forests with other much larger species of *Plethodon*, so we also summarize our studies, in chapter 6, of interactions between red-backed salamanders and *P. glutinosus*. Many other species of *Plethodon* are located outside of the geographic distribution of red-backed salamanders, but our research focused just on red-backed salamanders, their interspecific interactions, and their intraspecific social organization.

1.3 THE PLOT OF OUR RESEARCH PROGRAM

The red-backed salamander expresses a suite of behavioral patterns that is extraordinarily complex for an amphibian species (Jaeger, Gillette, et al., 2002). This complexity of behavior has been studied since 1965 by R. G. Jaeger and his research group leading to the publication of 114 articles thus far. We hope, by establishing approaches and methodologies that successively reveal the complexity of behavioral patterns in this species (e.g., the diversity of "decision rules"; Krebs, 1978), other ethologists will explore whether other species of the caudate amphibians have evolved similarly. We make no attempt to review the entire literature concerning *P. cinereus* but instead summarize our group's research from 1965 to 2015 with only occasional reference to publications by others. However, chapters 2 through 9 end with a summary of recent research by others to broaden the scope of our research.

While conducting our research, we attempted to follow the philosophical guidelines (Aucoin et al., 2005) established by Popper (1959; "refutation" of hypotheses), Platt (1964; "strong inference"), and Lakatos (1970; the "scientific research programmes"). Popper advocated that, due to inductive reasoning, scientists cannot prove anything to be true (Hume, 1748 [reprinted in 1995]); one can only attempt to refute or falsify hypotheses and eliminate those false hypotheses. Popper proposed that science progresses by forming a priori hypotheses that are testable and refutable. If a given hypothesis is refuted, then it should be abandoned. A limitation to falsificationism is that a single falsification may not be enough to abandon a particular theory or research program. Given this limitation, Lakatos suggested using the scientific research programme, which consists of replacing an old theory or hypothesis with a new model that is theoretically progressive because it is an advance over the predecessor. According to Lakatos, the new theory is progressive if it enables one to predict more than the predecessor and is occasionally empirically progressive, in that an observation confirms this new prediction. In so doing, a scientist's research program progresses by refuting or falsifying the accepted theory, or failing to do so, as proposed by Popper.

Refutation of a null hypothesis does not mean that the one alternative hypothesis is true. Scientists must consider all reasonable alternative hypotheses. Platt (1964) proposed that a progressive research program formulates multiple hypotheses that compete with each other. He suggested devising a critical experiment where the outcome will reject at least one hypothesis and researchers can repeat the process until the conclusions are no longer open to alternative interpretations, thus yielding a strong inference. By making, and often rejecting, new subhypotheses or new hypotheses (and predictions), the process is recycled. Our investigations were composed of a progression of studies, leading from one topic to another related topic: observations led to alternative hypotheses that led to experiments, which then led to the next set of observations, alternative hypotheses, experiments, ad infinitum. We attempt to summarize our research in this book in more or less chronological sequence to show the progression of our hypotheses and experiments. However, we also group these experiments into related topics (chapters 2–10) to tell a coherent story about *P. cinereus*.

The first area of research concerned interspecific competition between the red-backed salamander and the Shenandoah salamander (*P. shenandoah*, later listed as a federally endangered species) in Shenandoah National Park, Virginia. *Plethodon cinereus* is widespread in deciduous and coniferous forests of northeastern North America, while *P. shenandoah* is restricted to only three north- or northwest-facing slopes in Shenandoah National Park (the Pinnacles, Stony Man Mountain, and Hawksbill Mountain in the Blue Ridge Mountains; Petranka, 1998). Because they are lungless amphibians, plethodontid salamanders must respire through the skin and so are susceptible to desiccation during dry weather. To avoid desiccation between rainfalls, members of both *P. cinereus* and *P. shenandoah* either descend into underground cavities (where prey may be limited; Fraser, 1976) or preferentially move under rocks and logs where patches of moisture remain for days after the last rainfall and where prey tend to aggregate (Jaeger, 1970, 1979).

The parapatric distribution of these two species suggested the hypothesis of interspecific competition, with *P. cinereus* competitively dominant to *P. shenandoah*. While the latter species inhabits only three, dry, talus slopes, *P. cinereus* surrounds each slope in deep, moist soil more suitable to survival of species of *Plethodon*. The research on this suspected interspecific competition lasted through the late 1960s and into the early 1980s, and it is reviewed in detail in chapter 2. However, this research suggested the hypothesis that

P. cinereus holds interspecific territories against *P. shenandoah*, which led to a detailed study of intraspecific territoriality within *P. cinereus*; these behavioral experiments are reviewed in chapter 3. To understand territoriality more fully, experiments also investigated foraging tactics within territories by *P. cinereus* (chapter 4), because food seemed to be a limited resource in territorial defense, and the glands and pheromonal cues used during territorial advertisement and social communication (chapter 5).

Our studies of intraspecific territoriality provided a wealth of information about (1) olfactory and (2) visual communication among red-backed salamanders, (3) threat and submissive postures by territorial residents and intruders, (4) combat tactics, (5) the costs and benefits of aggression, and (6) foraging tactics within territories. Therefore, we were finally ready to return to studies of interspecific territoriality between *P. cinereus* and *P. shenandoah* and between *P. cinereus* and other species of salamanders (chapter 6).

Our studies of territoriality led to observations of seemingly complex social interactions within *P. cinereus*, both intra- and intersexually. For example, observers were surprised to see females often squash the fecal pellets of males and sometimes carry a pellet balanced on their nasolabial grooves (Walls et al., 1989). (In plethodontid salamanders, a pair of grooves lead from the upper lip to the nares, and in the grooves, cilia move nonvolatile chemicals from the substrate into the olfactory chamber; see "nose tapping" in Jaeger, 1984, 1986.) This observation suggested that females sample the dietary quality of males via their squashed feces (Jaeger & Wise, 1991). Therefore, we began an extensive series of investigations concerning the intra- and intersexual social behavior of red-backed salamanders (chapter 7). Chapter 8 relates the predator–prey contests between *P. cinereus* and a snake.

The wealth of salamandrine social behavior that we observed then led to our last major area of research: the cognitive ecology (Dukas, 1998) of red-backed salamanders. This area, reviewed in chapter 9, includes studies of (1) numerical discrimination of prey, (2) individual recognition memory of conspecifics, (3) how exposure durations and separation intervals affect recognition memory (e.g., of territorial neighbors and intruders), (4) the use of olfactory and visual cues for recognition memory, and (5) whether individuals can recognize an initial individual after interacting with other conspecific distractors.

We end our review with a coda (chapter 10) that attempts to weave together the various behavioral patterns spanning from interspecific competition (chapter 2) to cognitive ecology (chapter 9). In particular, we stress the large amount of ambient information that red-backed salamanders are able to input and process in their neural systems. These processes lead to the complex behavioral responses to conspecifics, congeners, prey, and other environmental stimuli that we report here. These cognitive-behavioral areas of research would be fruitful studies for other behavioral ecologists to pursue. Table 1.1 provides short definitions of the abbreviations that we use throughout this book.

Table 1.1 DEFINITIONS OF ABBREVIATIONS USED IN THE TEXT

Behavior	Definition
ATR	All trunk raised. A threat posture
BITE	Biting. Can be either a brief nip or an extended hold
CLT	Cloacal tapping. Cloaca is touched to the substrate; pheromonal marking behavior
CT	Chin tapping. A male touches his mental hedonic gland to the substrate; pheromonal marking used during courtship
EDGE	Edge. Salamander walks along the perimeter of a chamber; withdrawal or submissive behavior
FLAT	Flat. Salamander lies completely flat on the substrate; a submissive behavior
FTR	Front of trunk raised. The anterior one-third of the body elevated above the substrate; a resting, alert, and foraging posture
LA	Look away. Looks directly away from the opponent; a submissive display
LT	Look toward. Looks directly toward the opponent; a threat display
MA	Move away. Moves directly away from the opponent; a submissive display
MLBS	Mountain Lake Biological Station. In the Appalachian Mountains of southwestern Virginia
MT	Move toward. Moves directly toward the opponent; a threat or preattack display
NLG	Nasolabial grooves. Paired grooves from the upper lip's cirri to the nares; chemodetection structures
NT	Nose tapping. NLG touched to the substrate; chemodetection behavior
PCP	Postcloacal press. The postcloacal gland on the tail touched to the substrate; pheromonal marking behavior
RHP	Resource holding potential. "The ability to persist and win a contest" (Courtene-Jones & Briffa, 2014)
SNP	Shenandoah National Park. Site of Hawksbill and Blackrock Mountains in northwestern Virginia's Blue Ridge Mountains
SVL	Snout-to-vent length. Measured from tip of the snout to posterior end of the cloaca
TL	Total length. Measured from tip of the snout to the tip of the tail
ULL	University of Louisiana at Lafayette. Site of many behavioral experiments

NOTE: Agonistic displays of *Plethodon cinereus* are thoroughly defined in section 3.4.

1.4 COMMENTS CONCERNING METHODOLOGY AND STATISTICAL PARADIGMS

Plethodon cinereus is an excellent species for behavioral studies in the laboratory, though not in its forested habitats where individuals are largely invisible except during rainy nights. Individuals are small and thus thrive in small, transparent, plastic chambers that are easily cleaned. Their substrates need be only damp coffee filters or filter paper (in Petri dishes; Fig. 1.3) or paper towels (in larger Nunc bioassay dishes), and *Drosophila* is a complete diet for this species as adults but not for neonates. Room temperatures of 12° to 18°C are best for both housing

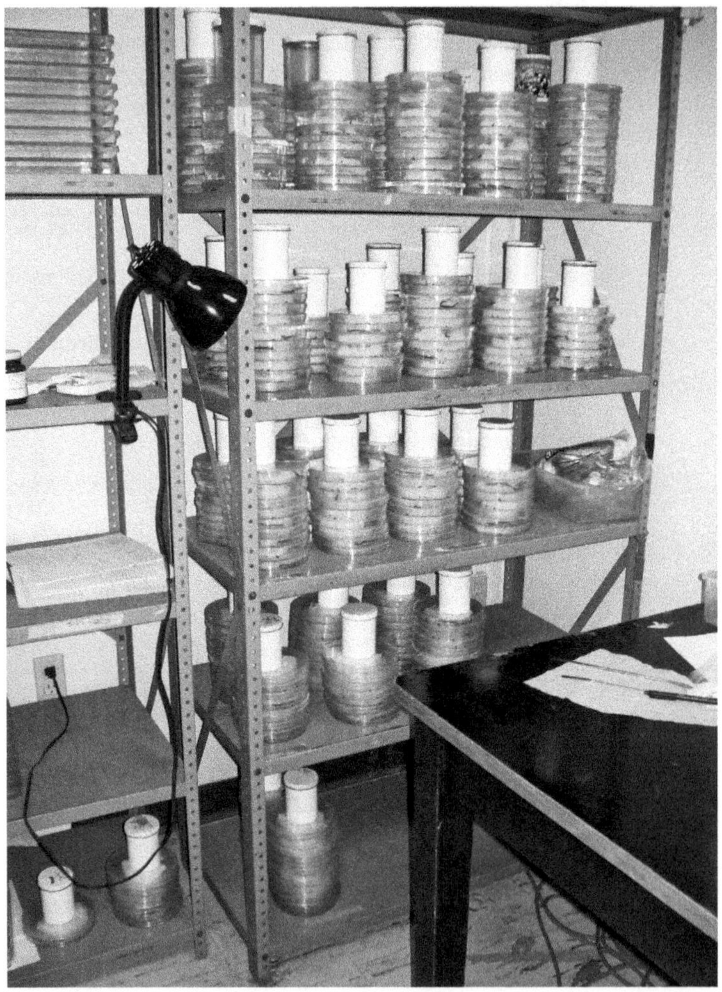

Figure 1.3 Stacks of single salamanders in their chambers (Petri dishes) in the laboratory at the University of Louisiana at Lafayette.

and behavioral testing. Salamanders in chambers, under dim light, readily feed and interact with each other while watched by slowly moving humans.

We have simultaneously housed up to 900 salamanders, either alone or in female–male pairs, in stacked chambers (Fig. 1.3) for up to 18 months with <1% mortality. For later behavioral experiments, the State of Virginia and Institutional Animal Care and Use Committee permitted us to collect large samples of *P. cinereus* at Mountain Lake Biological Station, transport them to the laboratory at the University of Louisiana at Lafayette, test them experimentally, and finally release them at a designated release site in the same forest. Therefore, salamanders were neither harmed nor killed. For experimental results to be meaningful, salamanders must be maintained in a healthy condition throughout the experiment: from the time that they are collected in the forest, transported to the laboratory, and maintained there. Jaeger (1992) and Buchanan and Jaeger (1995) provide advice for keeping salamanders in a healthy environment.

In the behavioral experiments reviewed here, nonparametric statistics were almost always used to analyze datasets instead of parametric, because our data often violated the assumptions of parametric statistics (e.g., analysis of variance and multivariate analysis of variance). Most data were grossly skewed (not normally distributed) and showed unequal variances. For example, in tests of time in all trunk raised (ATR), perhaps half of the tested animals would perform zero or little time in ATR while others would spend various amounts of time in ATR, yielding skewed results. Seaman and Jaeger (1990, pp. 337–340) warned of the dangers inherent in using parametric statistics for analyses of nonparametric data. Also, transformations of behavioral data, to meet the parametric assumptions, were almost never employed because Seaman and Jaeger (pp. 343–344) warned of flaws in using such a "philosophers' stone" to "correct" nonparametric data into parametric distributions. Ecological data were more often analyzed by parametric statistics when the data closely matched the statistical assumptions.

In many experiments reviewed here, nonparametric tests (e.g., the Wilcoxon-Mann-Whitney test, two-tailed; Siegel & Castellan, 1988) were used to compare the data from each salamander against itself under different test conditions. Comparing each individual against itself, among different conditions, greatly reduced the behavioral variances among individuals in the total sample size, thus reducing the chance of committing Type II errors in the statistical analyses. A priori power analyses were never used to determine an experiment's sample size, because early pilot studies and later experiments demonstrated that sample sizes of greater than 20 salamanders per test condition were sufficient to overcome behavioral variances found in *P. cinereus* and to yield statistical results that were usually either highly significant or highly nonsignificant.

Nonparametric statistics were also used for philosophical reasons, because they have lower power than parametric statistics (Siegel & Castellan, 1988).

Therefore, one's "favored hypothesis" (usually the alternative hypothesis), compared to the null hypothesis, is more likely *not* to be supported by a nonparametric test. One philosophy of science is that "scientific research programmes" (Lakatos, 1970) would be more progressive (yield "strong inferences"; Platt, 1964) when researchers attempt to reject their favored hypotheses (Popper, 1959). Continually failing, over many experiments, to reject a favored hypothesis adds credence that such a hypothesis may indeed be correct. We illustrate this philosophical concept with our reviews of the five experiments in chapter 4, which tested for optimal choice of diet, and the many experiments in chapter 3, which tested for territorial behavior by *P. cinereus*.

2

Interspecific competition between *P. cinereus* and *P. shenandoah*

The research program reviewed here began in 1965, after Highton (1962) published maps of the distributions of then known species of *Plethodon*. Of particular interest to Jaeger (1970) were the distributions of species of eastern, small *Plethodon*, in which "eastern" implies the eastern North American species and "small" identifies those species about the size of *Plethodon cinereus*. For examples, adults of *P. cinereus* reach total length (TL) of 65 to 125 mm (which varies among populations) while *P. glutinosus*, a "large" species, reaches TL of 115 to 205 mm (Petranka, 1998). Highton (1971) refined these maps to include newly discovered species.

These maps showed that *P. cinereus* has a vast distribution in forests from southern Canada to the mountains of northern North Carolina and from the Atlantic Coast to just west of the northern end of the Mississippi River (Petranka, 1998). By comparison, the other species of eastern small *Plethodon* have much more restricted distributions, mostly within the range of *P. cinereus*. Where *P. cinereus* comes into contact with another small species of *Plethodon*, their distributions often tend to be allopatric through much of the range with just narrow geographic areas of overlap (i.e., parapatry).

Parapatry could theoretically be caused by interactions between species. If two allopatric species live in different habitats separated by a narrow ecotonal boundary, then perhaps the two species are separated by preferences for those different habitats. For example, *P. nettingi* (80–110 mm TL) occurs in 68 small populations isolated on rocky mountain tops in the Cheat Mountains of West Virginia in red spruce–yellow birch forests (Petranka, 1998). Each population is completely surrounded by *P. cinereus*, which is found in deeper soil. One hypothesis is that this represents allopatry (and parapatry) due to differential habitat preferences.

Such habitat preferences, however, do not explain the parapatric boundary between *P. cinereus* and *P. hubrichti* (80–130 mm TL), which is closely related to *P. nettingi* (Petranka, 1998). *Plethodon hubrichti* occurs as a small population in a deciduous forest at Peaks of Otter in the Blue Ridge Mountains of Virginia, completely surrounded by *P. cinereus* in the same type of forest (Highton, 1971). One hypothesis is that interspecific competition between these two species causes this parapatric distribution, with the two species either in equilibrium or one species gradually replacing the other species (e.g., MacArthur, 1958, 1970; MacArthur & Levins, 1964). [N.B., Interspecific competition structuring the distributions of closely related species was a popular theory during the 1960s and 1970s.]

The question, then, was how to test among theories to explain the interactions between *P. cinereus* and other species of small *Plethodon*. This was the starting point for our 50-year research program from 1965 to 2015. A beginning occurred when Highton and Worthington (1967) discovered a new species, *P. shenandoah*, on three rocky mountain tops in Shenandoah National Park, in the Blue Ridge Mountains of Virginia. This is a small *Plethodon* (70–100 mm TL) and was originally named *P. richmondi shenandoah* but was later raised to a full species by Highton and Larson (1979).

2.1 ECOLOGICAL STUDIES

Jaeger (1970, 1971a, 1971b, 1972) chose Hawksbill Mountain as a convenient area of research, where the Appalachian Trail passes through areas with deep soil that are occupied by *P. cinereus*, through an area of parapatry, into in the drier Pleistocene-age talus where *P. shenandoah* occurs (Jaeger, 1970, Fig. 1). Jaeger's (1970) first task was to compose detailed maps of the chosen area of study, about 4 km upslope and 3 km across-slope (Jaeger, 1970, Figs. 1 and 2). These maps showed part of the large talus slope, including five isolated patches of talus farther downslope, the area of parapatry, and a section of deep soil inhabited by *P. cinereus*. Also, the areas of talus were surveyed to identify places containing Type I talus (bare rocks devoid of soil deposits), Type II talus (rocks with some soil and humus), and Type III talus (flat areas containing deeper soil) (Jaeger, 1970, Fig. 2).

The next task was to map the precise position of each salamander found: total $N = 822$; 61.7% *P. cinereus* and 38.3% *P. shenandoah* (see Fig. 2.1, as in Jaeger, 1970, Fig. 4). During 1966 and 1968, one 100 m^2 quadrat was censused for

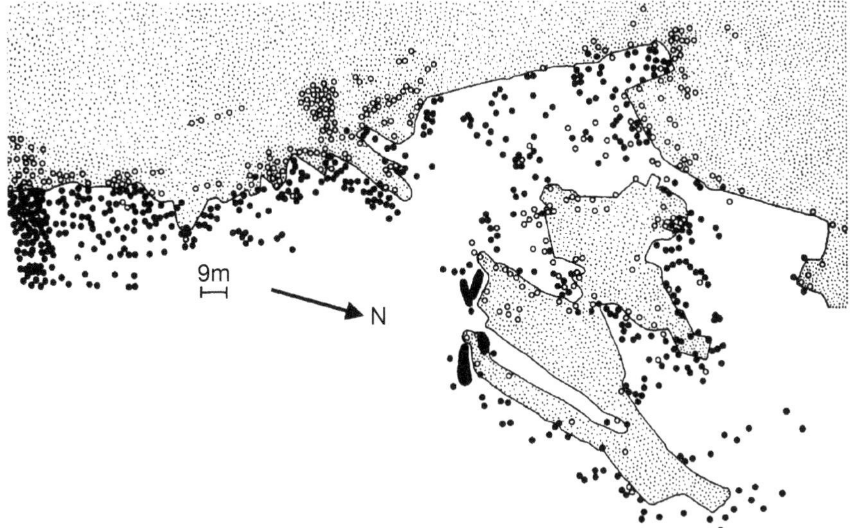

Figure 2.1 A portion of the talus–soil interface on Hawksbill Mountain. Solid lines indicate the interface, stippled areas are talus, and the white area is deep soil. Hollow circles (*P. shenandoah)* and solid circles (*P. cinereus)* represent the positions of all collected salamanders in this area. Large dark objects are large boulders. *Plethodon shenandoah* occurs in a metapopulation on Hawksbill Mountain, with the upper stippled area its source talus on top of the mountain and the lower stippled area one of its sink taluses downslope (see section 6.1 for details concerning this metapopulation). Reproduced with permission from the Society for the Study of Evolution: Jaeger, R. G. (1970). Potential extinction through competition between two species of terrestrial salamanders. *Evolution*, 24(3): Fig. 4, 637.

salamanders in each of the Type I, II, and III taluses and in deep soil 3 to 13 m outside the talus (Jaeger, 1970, Fig. 5). In both years, *P. cinereus* was abundant in areas with deep soil but absent in all three types of talus, while *P. shenandoah* was very rarely found in areas with deep soil, was abundant in Type III talus, was rarer in Type II, and was absent in Type I.

These distributions led to the formulation of three alternative hypotheses and one supposition, as explained here.

> H_1: *Plethodon shenandoah* competitively excludes *P. cinereus* from areas of talus.

This hypothesis was rejected (Jaeger, 1970) by censusing quadrates inside and outside of a large talus on nearby Blackrock Mountain. Neither *P. shenandoah* nor *P. cinereus* occurred in that talus, while *P. cinereus* was abundant in the surrounding deep soil (Jaeger, 1970, Fig. 5). These data provide evidence that *P. cinereus* does not occupy talus slopes in the absence of competitors. Thus the hypothesis that *P. cinereus* is competitively excluded from talus slopes by *P. shenandoah* was not supported.

> H_2: The two species are separated due to species-specific preferences for different habitats, with *P. cinereus* choosing deep, moist soil and *P. shenandoah* choosing drier talus slopes.

This hypothesis was rejected by Jaeger (1971b). Four laboratory experiments showed the following:

1. When given choices between rock and soil substrates, neither species expressed a preference when both substrates were moist (simulating rainfall), but
2. Both species significantly chose soil over rock as the substrates dried (simulating periods between rainfalls), as soil holds moisture longer. Therefore, for both species, substrate preference is based on moisture content and not on substrate texture.
3. *Plethodon cinereus* suffered significantly greater mortality and dehydration when subjected to drying rock compared to drying soil substrates. This suggested that the talus is a stressful environment for this species.
4. A dehydration experiment revealed that, compared to *P. cinereus*, the Shenandoah salamander survived longer and dehydrated significantly less per hour. This may explain why *P. shenandoah* has survived in talus slopes while *P. cinereus* is unable to enter far into the talus.

> H_3: *Plethodon cinereus* competitively excludes *P. shenandoah* from the deep soil habitat outside of the talus.

Jaeger (1971a) conducted an experiment on Hawksbill Mountain that provided support for this hypothesis. Three enclosures were sunk into the substrates in each of the Type I, II, and III taluses and into the deep soil outside of the talus. In each habitat, only *P. cinereus* ($N = 10$) was added to one enclosure, only *P. shenandoah* ($N = 10$) to the other enclosure (both control treatments), and an equal number of both species ($N = 10$ each) to the third, doubly large enclosure (the experimental treatments).

In the Type I enclosures, all salamanders died within one week, suggesting that this dry, rocky habitat is mortal to salamanders. In the Type II talus, over the 12-week experiment, *P. shenandoah* survived better than *P. cinereus* in both single species' and mixed species' enclosures. Therefore, as in Jaeger (1971b), the former species demonstrated longer survivorship under drying (i.e., shallow soil) conditions. In the Type III talus, both species survived equally well in the control treatments, while more *P. cinereus* survived longer in the mixed species treatment. Therefore, in the deeper soil within the talus, *P. cinereus* won the competitive race even though it does not naturally occur in these isolated pockets of soil where *P. shenandoah* is most commonly found.

In the deep soil enclosures (in the area of parapatry), both species had 100% survivorship in the control enclosures, but in the mixed species enclosure, *P. cinereus* had 90% survivorship and *P. shenandoah* only 20% survivorship over 12 weeks. This competitive release experiment suggested that *P. cinereus* is efficient in excluding *P. shenandoah* from areas of deep soil, but the experimental data were pseudoreplicated due to the design of only one enclosure per treatment (i.e., degree of freedom = 0). Therefore, this method of research (in the then-classic style of community ecology) had to be changed to a behavioral approach to answer three essential questions. (1) Is *P. cinereus* truly competitively superior to *P. shenandoah*? (2) If so, is the superiority due to exploitative competition (e.g., for food) or to interference competition? (3) If the latter, is competition due to interspecific territoriality? The answers are discussed here and in section 6.1, but first we present Jaeger's (1970) "supposition," as mentioned earlier.

The supposition was that "*shenandoah* is faced with potential extinction due to the erosion of soil into its talus refugium followed by a subsequent encroachment of *cinereus* and due to the paucity of isolated pockets of soil which are the centers of its distribution." Jaeger (1980a) provided unforeseen data that supported this supposition. A 14-year census (one census per year) of the population densities of both species on Hawksbill Mountain spanned a 28-day drought. The drought did not affect the density of *P. cinereus* in its deep soil habitat but, presumably, induced 99% mortality of *P. shenandoah* in its large talus habitat. In one Type III talus, *P. shenandoah* recovered to its predrought density in eight years. However, in an isolated, downslope patch of talus that lacked Type III habitats, the drought caused extinction of that population of *P. shenandoah*, and no reinvasion into this metapopulation (see section 6.1) was found over the last nine years of the census. Jaeger (1980a) concluded that *P. shenandoah* is an extinction-prone species.

The data in Jaeger (1970, 1980a) led to *P. shenandoah* being listed, in 1989, as a federally endangered species. Ironically, this severely limited future behavioral experiments concerning its interspecific competition for territories with *P. cinereus*, but see Griffis and Jaeger (1998) in section 6.1 as a major exception.

Because competition occurs when resources are limited (Jaeger, 1974a), Jaeger (1972) suggested that invertebrate prey are the limited resource that largely underlie competition between these two species of *Plethodon*. All age classes of *P. shenandoah* were found <3 m from the talus' edge in the parapatric area of deep soil, but only Shenandoah salamanders as large as or larger than the largest *P. cinereus* (in snout-to-vent length and head width) were found >3 m from the edge of the talus: head widths' mean = 6.3 mm for *P. shenandoah* and mean = 5.0 mm for *P. cinereus*. Based on stomach contents, both species ingested the same taxa of invertebrates, but >3 m from the talus, *P. shenandoah* ate prey significantly larger than those eaten by the cohabiting *P. cinereus*. These data suggested that *P. cinereus* competes strongly for prey in the first 3 m away from the talus, such that only the largest Shenandoah salamanders can move >3 m away from the talus. Prey become a limited resource periodically as availability of prey in the leaf litter is moisture-dependent. During and shortly after rainfalls, *Plethodon* can forage freely in the wet leaf litter. As the leaf litter dries, however, salamanders are confined to moist areas under rocks and logs, where prey are in limited supply (Jaeger, 1972, 1979).

Jaeger (1972) originally hypothesized that *P. cinereus* was competitively superior to *P. shenandoah* by exploitative competition for prey. Exploitation occurs when one species is more efficient in obtaining a share of a limited resource even though both species have free access to that resource. Later, Jaeger (1974b) posed the alternative hypothesis of interference competition. Interference competition occurs when the behavior of individuals of one species inhibits individuals of the competitor species access to a vital resource (such as prey) by interspecific aggression and perhaps by territoriality (Jaeger, 1974a). A laboratory experiment (Jaeger, 1974b) allowed the two species to compete for ten burrows in 30 cm of moist, deep soil in both single-species chambers (10 individuals: controls) and mixed-species chambers (5 of each species: experimental treatments). The resulting distribution of salamanders in burrows in control versus experimental chambers suggested some type of competition, due to shifts in niche breadths, which could have been caused by either exploitative or interference competition (or both) for access to burrows. Niche breadths were calculated by the depth that each salamander was found in the burrows: top litter and top, middle, or bottom soil.

2.2 BEHAVIORAL EXPERIMENTS

To solve this quandary, a series of behavioral experiments in the laboratory were begun to test both the exploitation and interference hypotheses. Kaplan (1977) examined capture rates of prey (*Drosophila melanogaster*) by size-matched *P. cinereus* and *P. shenandoah* ($N = 26$) in both single- and mixed-species tests. When tested alone, *P. cinereus* had a significantly faster rate of prey capture than its congener, but when the two species were tested together, *P. shenandoah* was significantly superior in rate of prey capture! Thus Kaplan (1977) concluded that differential exploitation of prey capture was insufficient to explain competitive exclusion of *P. shenandoah* from the deep soil on Hawksbill Mountain.

Bobka et al. (1981) again examined the exploitation hypothesis by determining the digestive assimilation efficiency of *P. cinereus* ($N = 20$) and *P. shenandoah* ($N = 12$), using C_{14}-labeled *D. melanogaster* as prey. No significant differences were found for assimilation efficiency of the species at either of the two ambient temperatures (10° and 20°C); however, the assimilation efficiency of both species was inversely related to temperature, suggesting that these salamanders may have difficulty maintaining a positive energy budget during warm weather in their natural habitats. Thus neither Kaplan (1977) nor Bobka et al. (1981) found evidence to support the exploitation hypothesis for the competitive advantage of *P. cinereus* on Hawksbill Mountain.

Wrobel et al. (1980) tested the interference hypothesis in a laboratory experiment where pairs of *Plethodon* were placed in chambers and monitored over 58 days. The replicated pairs were randomly divided among the (1) interspecific experimental group, *P. cinereus* and *P. shenandoah* ($N = 13$ pairs), (2) the *P. cinereus* only control group ($N = 13$ pairs), and (3) the *P. shenandoah* only control ($N = 13$ pairs). They placed a limited number of *D. melanogaster* in each chamber such that if the two salamanders shared the flies equally, both would lose body mass over 58 days. Salamanders were fed every three days, and behavioral interactions between each pair of salamanders were recorded. "Winners" and "losers" were declared at the end of the experiment where the winning salamander gained more or lost less mass than the cohabiting loser. This experiment was designed to induce competition for food (as on Hawksbill Mountain) between pairs of salamanders and to learn if intra- and interspecific aggression explained winners and losers. During the experiment the authors recorded (1) approach (move toward: a potential threat behavior), (2) body contact without a bite, (3) head attack (bite, snap, and tongue flick), and (4) withdrawal (move away from the "aggressor"). Both species expressed intraspecific aggression, with males showing significantly more aggression than females. For interspecific pairs, winners exhibited significantly more aggressive activity than losers, which suggested that aggressive superiority and foraging superiority are associated (e.g., in the area of parapatry on Hawksbill Mountain). *Plethodon cinereus*, however, did not win significantly more often than *P. shenandoah*, although the former was more aggressive on average.

Wrobel et al. (1980) concluded that interference competition occurs both within and between the two species. This raised the question as to whether territorial aggression is involved in intraspecific and interspecific conflicts, which we discuss in detail in chapters 3 and 6, respectively. The data in Wrobel et al. (1980), however, still did not answer why or how *P. cinereus* can behaviorally dominate *P. shenandoah* in the area of parapatry outside the talus on Hawksbill Mountain. This problem still confounds us today, 45 years after Jaeger (1970) was published.

Jaeger and Gergits (1979) also explored the hypotheses of intra- and interspecific territoriality in red-backed and Shenandoah salamanders by examining possible pheromonal (within species) and allomonal (between species) communication through chemical signals on the substrate. Their rationale was that territorial advertisement in *Plethodon* would be by olfaction, because individuals in this genus have many integumental glands but lack external ears (for vocal communication). Additionally, in this species, visual communication might be very limited in its complex leaf litter habitat. In this laboratory experiment, Jaeger and Gergits tested 200 adult *P. cinereus* and 100 adult *P. shenandoah* for 300 minutes each to learn if they would choose (in experimental treatments) to move toward their own previously marked substrate on one side of a chamber or toward another salamander's previously marked substrate on the other side of the chamber. They also conducted control treatments with a salamander's own substrate on both sides of the chamber. During intraspecific experimental treatments for *P. cinereus*, males were confronted with their own marked substrates versus some other male or female substrates and females were confronted likewise. During interspecific experimental treatments, males and females of each species were confronted with their own substrates versus substrates of male or female congeners.

For the intraspecific experiments, males and females of *P. cinereus* showed significant preferences for their own substrates only when simultaneously exposed to the substrates of conspecific males. (An intraspecific experiment was not performed with *P. shenandoah* because *P. cinereus* is the dominant species on Hawksbill Mountain.) These results suggested that (1) movements and subsequent substrate choices of *P. cinereus* are influenced by the identity of the salamanders previously inhabiting the substrates; (2) consequently, these salamanders leave pheromonal signal markers on substrates where they reside; and (3) these pheromonal signals contain information about the sex of the previous inhabitant.

For the interspecific experiments, both *P. cinereus* and *P. shenandoah* significantly avoided substrates of all congeners, except that females of *P. cinereus* did not avoid substrates of male *P. shenandoah*. These results suggested that (1) movements and substrate choices of both species are governed by the identity of the salamanders previously inhabiting the substrates, therefore, (2) both species apparently leave allomonal signal markers on the substrates upon which they move, and (3) these signals contain information as to the species and, perhaps, the sex of the marker.

Overall, the important conclusion from Jaeger and Gergits (1979) was that, for both species, glandular secretions and/or fecal/urinary materials act as signal markers that could be used in territorial defense within *P. cinereus* and between the two congeners.

2.3 SELECTED RECENT RESEARCH BY OTHERS: INTERSPECIFIC COMPETITION

In their extensive studies of *P. cinereus* and *P. shenandoah*, Jaeger and colleagues provided a model system for exploring the intersection of animal behavior and ecology. Since that time, additional techniques such as morphometric and ecological modeling have become available to ecologists. For example, Dean Adams (Iowa State University) and his students established that, in some species of *Plethodon*, the evolution of head shape correlated well with diet. Thus the study of head shape presents one way to draw inferences about competition for prey. This approach is well suited to asking questions about the nature of competition between species that overlap only partially in geographic range such as *P. cinereus* and *P. shenandoah*. Myers and Adams (2008) examined specimens of *P. cinereus* ($N = 522$) and of *P. shenandoah* ($N = 236$) from three sympatric localities and 11 allopatric localities where only *P. cinereus* occurs. They examined 18 shape variables associated with cranial morphology and found significant variation in shape across geographic space. Specifically, they found that upland populations of *P. cinereus* differed from lowland populations, a result mirrored in a previous study of *P. cinereus* in New York (Maerz et al., 2005). They did not, however, find any evidence that shape differences between *P. cinereus* and *P. shenandoah* were heightened in areas where the two species interact. This result is interesting given the strong evidence from behavioral and ecological studies suggesting that the two species compete. Instead, they argued that the presence of refugia for *P. shenandoah* (the talus habitat) might serve to decouple morphological evolution from competition. Under this scenario, there is no selective pressure forcing either species to diverge on a prey niche dimension. Alternatively, they suggested that morphological evolution might be constrained in this system due to limited genetic variation on the part of *P. shenandoah* or limitations on the types of prey available in the shared habitat.

Other approaches to studying competition in *Plethodon* have utilized the vast museum holdings of specimens to examine patterns in community composition. The aim of this approach is to determine whether communities of *Plethodon* are assembled randomly or whether the interactive forces of competition can be detected. In one such analysis, Adams (2007) examined 45 species of *Plethodon* from over 4,500 geographic localities for patterns of species co-occurrence. Adams found that species of similar body size were less likely to co-occur within communities. He argued that his results, plus the experimental evidence from species-pair studies such as between *P. cinereus* and *P. shenandoah*, strongly indicate that community composition in this genus emerges from competitive interactions among species.

The use of geographic information system technology and niche modeling can be used to ask questions regarding expected versus observed geographic ranges and to tease apart the relative roles of physiological tolerances of species and behavioral interactions between species. For example, Arif et al. (2007) examined the competitive interactions between *P. cinereus* and *P. hubrichti*. Like *P. shenandoah, P. hubrichti* is a mountaintop species whose range is surrounded by *P. cinereus*. Unlike *P. shenandoah*, it has no habitat refugium where it can avoid interactions with *P. cinereus*. Using a combination of bioclimatic modeling and behavioral tests, they found that *P. cinereus* is excluded from montane areas due to competition with *P. hubrichti*, while *P. hubrichti* is restricted to those same areas due to climatic factors. Their results were further highlighted by studies using mechanistic niche models that link functional traits of the species of interest to bioclimatic data. Gifford and Kozak (2012) applied this approach to *P. jordani* and *P. teyahalee* to illustrate how montane species are rarely limited in their downslope movement by competition but are instead limited by physiological factors related to climate. These two approaches have not yet been applied to *P. cinereus* and *P. shenandoah*, but they present exciting avenues for future research.

The vulnerable status of *P. shenandoah* is a concern to conservation biologists. Potential threats to the species include genetic introgression with *P. cinereus*, loss of genetic diversity due to small population sizes, and habitat alteration via climate change. Carpenter et al. (2001) examined the population genetics of *P. shenandoah* and *P. cinereus* within Shenandoah National Park and found no evidence of introgression or loss of genetic diversity. These results suggest that *P. shenandoah* is as demographically stable as *P. cinereus* and is at no risk of extinction due to hybridization with its widespread congener. It is unknown whether human-caused changes in climate will accelerate the loss of this species, but niche modeling of other montane species suggests that even small changes in climate may threaten the ranges of many high-elevation species (Gifford & Kozak, 2012). We now continue to chapter 3, which discusses the costs, benefits, and the innumerable ambient variables that shape territoriality in red-backed salamanders in Virginia.

3

Intraspecific territoriality by *P. cinereus*

3.1 DEFINITION AND THEORY

Chapter 2 ended with a strong suspicion that *P. cinereus* is territorial, and in this chapter we discuss the many studies involved in determining that red-backed salamanders are territorial intraspecifically and that many biotic and abiotic variables influence the outcome of contests between rivals. These studies led to later research, summarized in section 6.1, on how *P. cinereus* holds territories against *P. shenandoah*. We begin, however, with a definition of a territory and some underlying theories that were then pertinent to our research.

We define a territory as an exclusive area that is defended against rivals by individuals or groups (Brown & Orians, 1970; Davies & Houston, 1984). According to Davies (1978), territories are evident whenever individuals or groups of animals are uniformly spaced more than would be expected from random occupation of suitable habitats. Territories are usually maintained by defense/aggression and may be advertised by more subtle signals such as chemical cues or scent marks (*sensu* Gosling, 1990), vocalizations, or displays (Davies, 1978; Davies & Houston, 1984). The behavioral investment in territorial defense must theoretically provide, on average, advantages to an individual's overall fitness (Brown & Orians, 1970). Territorial behavior should generally occur when the benefits ("payoffs") of obtaining the territorial resource (e.g., feeding area, mate, home range) outweigh the potential costs of defense (Maynard Smith & Parker, 1976). Potential costs of defense include increases in energy spent advertising and defending the territory, increases in time spent fighting, and increases in injury during escalated contests (Davies, 1978; Hardouin et al., 2006; Jaeger, 1981b). However, most contests over territories are asymmetric in nature, and these asymmetries may settle contests without escalation (Maynard Smith & Parker, 1976). Contests can vary in payoff asymmetries where one individual may have more to gain by winning the contest, such as a territorial owner who has invested more time and energy into exploration and settling boundaries with neighbors (Maynard Smith & Parker, 1976). Contests can also vary in resource holding potential (RHP) where individuals may differ in some intrinsic feature such as body size (Maynard Smith & Parker, 1976). The experiments reviewed here used many of these theories to design and explore facets of territorial behavior of *P. cinereus.*

Based on observations of aggression between and within species of *Plethodon*, Thurow (1976) was the first to suggest that *P. cinereus* is territorial. However, in order to determine if territoriality actually occurs in *P. cinereus*, Jaeger and Gergits (1979) stated that adult individuals of *P. cinereus* should (1) exhibit site tenacity, (2) aggressively defend the area, (3) advertise the defended area, and (4) often be able to expel potential competitors. Jaeger and Gergits set these rigorous criteria because *P. cinereus* is nearly impossible to observe behaviorally in its forested leaf litter habitat, so tests of territoriality required behavioral experiments in the laboratory and ecological experiments in the forest.

3.2 DISTRIBUTION AND PREY AVAILABILITY

In the Blue Ridge Mountains of Virginia, *P. cinereus* can be seen climbing plants on rainy and foggy nights. Salamanders start climbing plants just after twilight, where they feed on aphids, spiders, snails, and other invertebrates, then descend to the forest floor around 2 AM (Jaeger, 1978). Experiments conducted by Fraser (1976) and observations by Jaeger (1972) suggested that food is a limited resource for *P. cinereus* and that even during wet periods, when food is more abundant, prey may not be available on the forest floor. This led Jaeger (1978) to ask two questions: (1) Why do salamanders forage on plant-dwelling prey when small invertebrates appear to be abundant in the forest leaf litter? (2) Why do the salamanders spend only a few hours climbing plants during wet nights? He tested two hypotheses to address these questions:

> H_1: Climbing plants allows salamanders to exploit a food supply that is more readily available to them than prey on the forest floor.

H_2: Even during rainy nights, salamanders can lose body water to the air, which could lead to desiccation just after a few hours of exposure; exposure should be more severe on plants than in the leaf litter.

On Blackrock Mountain (near Hawksbill Mountain), Shenandoah National Park (SNP), Jaeger (1978) searched 300 m^3 plots of ground and plants for 1 hour during seven dry nights, when no rain had occurred within 8 hours of the observations, and eight wet nights, when there was ≥1 cm of rainfall up to 8 hours before the observations. Salamanders climbed plants only on rainy nights or when it had rained in the afternoon of the observations. Thus salamanders were significantly more active in climbing plants on wet nights compared to dry nights.

Next, Jaeger (1978) collected salamanders from under cover objects during the day and on plants or in the leaf litter at night on two different days ($N = 35$ salamanders, except on one night, $N = 21$). He preserved the salamanders before measuring each one for snout-to-vent length (SVL) and estimating total prey volume in each salamander's stomach. Those salamanders climbing plants for 1 hour, at night, had a significantly greater volume of food in their stomachs than conspecifics foraging concurrently in the leaf litter. He then determined the families of prey found in the stomach of each salamander. Homopterans and hemipterans made up 13.2% by volume of the total diet of plant climbing *P. cinereus*. This was 6.5 times the volume of homopterans and hemipterans found in the stomachs of salamanders collected during the day or in the stomachs of salamanders collected at night in the leaf litter. Thus salamanders consumed plant-dwelling insects while climbing trees and other plants, and these insects tended to be larger than the mites and collembolans that composed a large fraction of salamanders' diet in the leaf litter. These results supported the inferences in chapter 2 that *P. cinereus* is food limited in the leaf litter, which might induce intra- and perhaps interspecific competition for food at Jaeger's research areas in Virginia. Such food limitation is evidenced by *P. cinereus* climbing plants on rainy nights and ingesting more prey there than in the leaf litter: that is, they leave the leaf litter and find more and larger prey on plants.

To answer Jaeger's (1978) second question regarding why the salamanders spend only a few hours climbing plants during wet nights, he measured the extent of desiccation of salamanders placed in screen cages on plants during wet and dry nights. First, 20 salamanders were allowed to become fully saturated over a 24-hour period. Then half of those salamanders' cages were placed on wet leaves and suspended from plants (experimental group). The other half were placed on wet paper towels, and their cages were watered hourly overnight while suspended from plants (control group). The next morning, he measured the weight change of each salamander through water loss or gain. Both wet and dry night experimental groups lost more body water than controls, and the dry night experimental group lost significantly more water than the wet night experimental group. Therefore, Jaeger concluded that salamanders spend only a few hours climbing plants during wet nights as a mechanism for avoiding desiccation.

Salamanders also avoid desiccation on the forest floor during dry periods by remaining in patches of moisture under rocks and logs (Fraser, 1976). This restriction in movement also decreases their chance of finding prey, which are more abundant in the leaf litter (Fraser, 1976). Together these data suggest that prey are not necessarily limited in abundance (i.e., in the leaf litter) but are not accessible when the salamanders move under rocks and logs. This distinction is important for understanding both intra- and interspecific competition for food by *P. cinereus* (Jaeger, 1972).

Up until this point in time, it was unclear what microhabitats individuals of *P. cinereus* used on the forest floor. Jaeger (1972) suggested that salamanders move horizontally, spending time in the leaf litter during wet periods while spending time under cover objects during dry periods. If this were the case, then the surface density of salamanders should not vary much during dry and wet periods. Fraser (1976) suggested a vertical movement pattern of salamanders, where salamanders spend time in the leaf litter during wet periods and move underground during dry periods. If this were the case, then the surface density of salamanders should decrease with decreasing amounts of rainfall.

At Blackrock Mountain, Jaeger (1980c) sampled the surface density and microhabitats of *P. cinereus* on the forest floor for 22 sampling days during the spring and summer, which included a 7-day period without rainfall. He set up 22 nonoverlapping sample sites (100 × 100 m) that he sampled every other day. Within each sampling site, salamander density was determined by using a 1 m^2 quadrate frame placed on the ground at a randomly chosen distance from the center of the plot in one of eight randomly selected compass directions. In each quadrate, he turned over every leaf, twig, rock, and log. He then counted the number of salamanders found and recorded the microhabitat (soil-leaf litter temperatures, salamander position, soil depth, leaf litter depth, surface area of rocks, and daily rainfall). He found that the surface density of salamanders did not vary significantly over the 22-day period. We now infer that these data support Jaeger's (1972) hypothesis that *P. cinereus* moves linearly between the leaf litter and cover objects during wet and moist periods but also support Fraser's (1976) idea that *P. cinereus* moves vertically underground during extremely dry periods, because Jaeger (1980c) also found that the salamanders' surface population decreased during long rainless periods. Thus the percentage of surface salamanders in the forest leaf litter increased and the percentage of salamanders under rocks and logs decreased with increasing amounts of rainfall. Horizontal movement on the forest floor did not usually lead to aggregations of salamanders under the same rocks and logs, at least during the summer noncourtship season in Virginia (Jaeger, 1979). This may be the result of interference competition for space and food availability and hints at territoriality in *P. cinereus* (Jaeger, 1979; Jaeger & Gergits, 1979; Thurow, 1976).

Jaeger (1980b) tested the hypothesis that food is a limiting resource for *P. cinereus*. Using salamanders from Blackrock Mountain, he conducted a study to provide information on the foraging success of *P. cinereus* in its natural habitat under various moisture levels and temperatures. Also, he conducted a laboratory study that yielded estimates of the energetic requirements (i.e., via ingested prey) of *P. cinereus* at various temperatures. On the forest floor, he set up 15 quadrats (15 × 100 m) that he sampled at random, one each day for 15 days. During each sample day, he collected 30 to 45 individuals that were measured for SVL and weight, and the prey, from each stomach, were identified ($N = 9{,}025$ prey) and measured and their volume determined. Thus he determined the total number and volume of prey ingested in a day by each of 1,073 salamanders.

Jaeger (1980b) took another 52 adult salamanders into the laboratory to test the volume of food needed per day to maintain a constant body weight. The salamanders were randomly partitioned into four groups ($N = 13$ each) that were fed 0, 1, 4, or 8 *Drosophila* per day for 28 days. He weighed the salamanders at the beginning of the experiment, then weekly, at three different temperatures (6.6°C, 10.0°C, and 15.5°C) during three different experiments. He determined the mean volume of food ingested per mm of salamander per day, the percentage weight change of each salamander, the metabolic requirement for the three temperatures (from Merchant, 1970), and whether each salamander was on a positive or negative energy budget. A negative energy budget occurred when the calories ingested per day were less than the calories metabolized per day (i.e., resulted in weight loss). Jaeger also estimated the caloric content of food in each salamander, and food assimilation at various temperatures, to determine the energy input for each salamander per day (from Bobka et al., 1981). By comparing his results from the laboratory with those from Blackrock Mountain, he found that the foraging success of *P. cinereus* in the forest was directly and significantly related to rainfall and that temperature determines the metabolic requirements and food assimilation efficiencies of those salamanders. He concluded that on most but not all days, salamanders in the forest were on a negative energy budget and prey were a "limiting" resource, especially during dry, warm days. Therefore, he inferred that salamanders on the forest floor should compete for food that is periodically limited in availability.

3.3 SITE TENACITY BY *P. CINEREUS*

According to Jaeger and Gergits (1979), one of the four variables needed for territoriality to occur is site tenacity, and previous research indicates that *P. cinereus* has small home ranges of about 3.0 to 4.8 m^2 (Kleeberger & Werner,

1982; Taub, 1961). This led Gergits and Jaeger (1990b) to ask the question: How faithful are salamanders to a given site in their natural habitat? They hypothesized that *P. cinereus* exhibits true site attachment; that is, an individual confines most of its activities and time, while on the surface of the forest floor, in the limited area under or near a certain cover object (rock or log). At Blackrock Mountain, Gergits and Jaeger set up a 5 × 5 m quadrat on the forest floor where they flipped over cover objects and leaf litter during the day in spring, summer, and autumn to census the number and location of adult (>28 mm SVL) salamanders. Sampling occurred only on the day following rainfall ($N = 12$ days) to avoid frequent flipping of cover objects and destroying the leaf litter habitat. They noted the original capture site of each salamander ($N = 90$), the subsequent exact recapture site, and the distance between each recapture site and the original capture site for each salamander. Of the 55 individuals recaptured, 37 salamanders were recaptured once, 12 salamanders twice, 4 salamanders three times, 1 salamander four times, and 1 salamander five times. These recaptures occurred under 50 cover objects, and 91% were within 1 m of the original capture site. Gergits and Jaeger found few recaptures in the leaf litter, no recaptures outside the quadrat, and no significant difference between recaptures within the same season and recaptures between seasons. Therefore, they inferred that *P. cinereus* has extreme site (cover object) attachment and that this site attachment occurs across seasons.

3.4 DETERMINING SEX AND DEFINING BEHAVIORAL PATTERNS

Before we continue our discussion of whether *P. cinereus* is territorial, we must first define and give full descriptions of the behavioral patterns recorded during the numerous laboratory experiments found in this chapter and in the remainder of this book. In this section, we also describe how to determine the sex of red-backed salamanders.

Plethodon cinereus, and some other plethodontid salamanders, can be sexed by a method called candling (Gillette & Peterson, 2001). An individual is placed into a clear plastic bag, held up to a light source (e.g., sunlight or lamp), gently squeezed so that the light can pass through its translucent skin, and then checked in the abdominal area for the presence or absence of dark testes. During the courtship season, males also have square snouts, mental glands (a whitish area on the chin), and a whitish and slightly enlarged region immediately lateral to the cloaca. Females have rounded snouts and no whitish region around the cloaca and lack mental glands, and, if they are gravid, ova are visible through the abdominal wall (Gillette & Peterson, 2001).

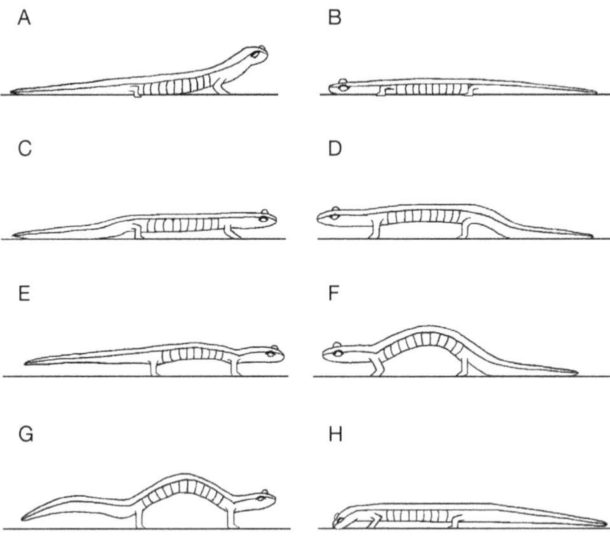

Figure 3.1 Behavioral postures used by *P. cinereus* during intra- and interspecific interactions and chemoinvestigation. (A) FTR = front of trunk raised, a resting or alert posture. (B) FLAT = a submissive posture. (C) ATR 1 = all of trunk raised, a low-threat posture but also used while foraging and moving. (D) ATR 2 = an escalated threat posture. (E) ATR 3 and (F) ATR 4 = more severe threat postures. (G) ATR 5 = maximal threat posture, often proceeding a biting attack. (H) NT = nose tap, a chemodetection behavior, often used when monitoring pheromones and allomones. Drawing by Nancy Kohn.

Jaeger (1981b, 1984) described many of the agonistic behaviors of *P. cinereus*. Here we summarize the behavioral patterns recorded (also see Fig. 3.1) during various experiments.

Front of trunk raised (FTR)—The front legs are extended downward so that the head and anterior portion of the trunk are raised off the ground. This is a resting and alert posture (Fig. 3.1A).

All trunk raised (ATR)—A gradational series of postures (Jaeger & Schwarz, 1991) that signal increasing levels of aggression starting with ATR 1 (Fig. 3.1C) where the individual rises from a resting posture into a low stance. The legs are not fully extended, the back is unarched, and the tail is resting on the substrate; this is also used when rapidly walking and pursuit-foraging. In ATR 2 (Fig. 3.1D), the limbs are fully extended downward until the salamander is in a high stance, with an unarched back and its tail on the substrate. The tail can then be raised off the substrate with the back not arched (ATR 3; Fig. 3.1E), or the back can be arched with the tail remaining on the substrate (ATR 4;

Fig. 3.1F). The most aggressive posture is a high stance with the back arched and the tail held off the substrate (ATR 5; Fig. 3.1G), which is considered to be a "look big" threat posture. ATR 5 often precedes a biting attack.

Flattened (FLAT)—The entire body is pressed against the substrate in a submissive posture (Fig. 3.1B).

Nose tapping (NT)—The head is lowered (~45°) to the substrate such that the nasolabial cirri come into contact with the substrate. This is used in chemoinvestigation (Fig. 3.1H).

Chin tapping (CT)—From ATR or FTR, the head is lowered until the chin touches or scrapes along the substrate.

Cloacal tapping (CLT)—From ATR, the pelvic area is lowered until the cloaca contacts the substrate. This is used in marking the substrate.

Biting (BITE)—An aggressive behavior whereby one salamander grasps another salamander with its mouth. Bites occur infrequently, but when they do they are usually directed toward the tail or nasolabial grooves (NLG) on the snout (Jaeger, 1981b). Bites can be either fast "nips" that do not penetrate the opponent's skin or extended "holds" that may lead to injury to the skin.

Edging (EDGE)—When a salamander roams the periphery of a test chamber pressing its snout against the wall or crevices between the lid and the walls of the chamber (Wise & Jaeger, 1998). This is an "escape" behavior and is considered submissive.

Look toward (LT)—A salamander turns its head in the direction of a conspecific as if looking at it; an aggressive behavior.

Look away (LA)—A salamander turns its head such as to avoid visual contact with a conspecific; a submissive behavior.

Move toward (MT)—One salamander moves directly toward the body of another salamander; an aggressive or preattack behavior.

Move away (MA)—One salamander moves directly away from the body of another salamander; a submissive behavior.

Touching (TOUCH)—In which any portion of one salamander's body (exclusive of bite) contacts the body of another salamander (Gillette, Jaeger, et al., 2000).

3.5 THE USE OF ODORS AND DEAR ENEMY RECOGNITION

Salamanders have many glands that are involved in communication (see chapter 5), and these glands may be used to mark substrates in their home ranges. Tristram (1977) investigated the ability of individuals of *P. cinereus* to distinguish their own odors from those of conspecifics. She collected salamanders from Blackrock Mountain and housed them individually in the laboratory; they remained in individual containers for 14 days, covering the substrates (filter paper) with their own secretions and waste. On day 15, she tested the salamanders in three conditions (C): the response of salamanders to substrates (filter paper) with C_1: their own odor ($N = 14$), C_2: the odor of unfamiliar conspecifics ($N = 28$), and C_3: no salamander odor as controls ($N = 14$). She measured the number and duration of NT (a chemoinvestigative behavior; see Fig. 3.1H and chapter 5). She found a significant difference in the number of NT between the substrates with the salamanders' own odor versus the conspecific odors and a significant difference between substrates with their own odors versus the controls. There was no significant difference in the number of NT on substrates with controls and conspecific odors. Thus the number of NT was much greater on substrates with self-marked odors than either conspecific odors or controls. Tristram concluded that individuals could detect olfactory cues from their environment, which allows them to distinguish between their own substrate markings and those of conspecifics.

McGavin (1978) then tested the hypothesis that salamanders can learn to distinguish among odors of conspecifics; that is, they can recognize one animal's marking as distinct from those of others. She measured NT in order to determine whether individual *P. cinereus* can learn to recognize the substrate odor of a single conspecific and distinguish between familiar and unfamiliar odors. She defined "familiar" as substrate markings of a given conspecific that the experimental salamanders had prior exposure to and "unfamiliar" as conspecific's substrate markings that the salamander had never previously encountered. Adult salamanders collected from Albany County, New York, were allowed 14 days to become familiar with a paired conspecific. (Later, Kohn and Jaeger, 2009, found that 8 hours is sufficient time for familiarity to occur; see section 9.5.) On day 15, the salamanders were then tested for their responses to familiar conspecifics ($N = 11$) and unfamiliar conspecifics ($N = 13$). She found that adult salamanders were 2.5 times more likely to NT the substrates from familiar conspecifics than from unfamiliar conspecifics. Thus she inferred that *P. cinereus* can distinguish between substrate markings of familiar and unfamiliar conspecifics.

Previous research with *P. cinereus* suggested that adult salamanders tend to remain in a given area, under rocks and logs that are used for feeding while avoiding cover objects marked by others. This led Jaeger (1981b) to test for dear enemy recognition in red-backed salamanders. Dear enemy recognition is defined as showing more aggressive behavior toward strangers than toward territorial neighbors. This minimizes the energy expended on hostile acts (Wilson, 1975) by reducing aggressive escalations between neighbors, because once territorial boundaries are established between neighbors, they pose little threat to an individual's territory (Jaeger, 1981b). Also, time defending a territory reduces time for foraging, finding mates, and other activities (Davies, 1978; Golabek et al., 2012; Hardouin et al., 2006). An unfamiliar stranger, however, may be trying to establish (or invade) a territory, so more time and energy is needed to defend territories against strangers (Hardouin et al., 2006; Jaeger, 1981b; Palphramand & White, 2007). Escalated contests are also more likely to occur when the payoffs from winning a territory are large compared to the loss if injured (Maynard Smith & Parker, 1976). Thus an individual should (in theory) escalate a contest if it currently lacks a territory and could possibly win one or if the individual currently has a low quality territory and is seeking a better one. Escalation should seldom occur between neighbors where the payoff is small (Maynard Smith & Parker, 1976).

Jaeger (1981b) tested two hypotheses:

H_1: That *P. cinereus* employs dear enemy recognition using chemical signals to distinguish familiar from unfamiliar conspecifics.

H_2: That the cost of an escalated contest (biting) is greater for losers than for winners.

Using adult male and female *P. cinereus* from Albany County, New York, he conducted three experiments (E): E_1: a laboratory study where the salamanders had no prior residency in an area, E_2: a laboratory study where the salamanders had prior residency in an area, and E_3: a laboratory and forest study looking at the cost of losing a contest. In E_1, paired adults were placed into chambers (18 × 18 × 8.5 cm) that were divided into three sections by opaque sliding doors, where one salamander was placed into each of the end sections. For 24 hours, one of the salamanders was allowed to roam into the center area where it could deposit its pheromones. The first salamander was then placed back into its end chamber and the sliding door was closed so that there was no physical or visual contact between the two salamanders. For the next 24 hours, the second salamander was allowed into the center area. On days 7 and 8, each salamander was tested twice, in random order, in a neutral arena (32 × 17.5 × 1.5 cm) with a familiar salamander from the same chamber ($N = 82$) and with an unfamiliar salamander ($N = 82$) from a separate chamber. Jaeger (1981b) recorded visual displays (both ATR and FLAT; see Fig. 3.1) and the frequency of bites. He found significantly more bites between unfamiliar salamanders than between familiar salamanders, significantly more time in ATR toward unfamiliar conspecifics than familiar conspecifics, and significantly less time in FLAT toward unfamiliar conspecifics than familiar ones. Therefore, he inferred that unfamiliar salamanders threatened and bit more and assumed submissive postures less often than familiar ones. Also, there were no significant differences between males and females, indicating that both sexes are capable of defending areas (territories).

In E_2 he placed 23 salamanders ("residents") into separate chambers (31 × 16 × 9 cm) for six days with a surrogate (a rolled piece of damp paper towel about the size and shape of a salamander) containing either the pheromones of a conspecific ($N = 23$) or a surrogate that had no conspecific's pheromones. On day 7, in random order, a live familiar conspecific (in which the surrogate paper towel had this salamander's odors) or a live unfamiliar conspecific (the resident had no prior experience with odors) was introduced into the resident's chamber and biting was recorded. Jaeger (1981b) found that residents were more likely to initiate an aggressive attack than those salamanders introduced to the chambers. He also found significantly more biting by residents toward unfamiliar conspecifics than toward familiar conspecifics. These bites were directed at the snout, tail, or trunk with significantly more bites on the snout than on the trunk and on the snout than on the tail. Bites on the tail can lead to tail autotomy, which would result in the loss of fat reserves used during poor foraging periods and during (for females) reproduction. Bites on the snout could lead to scar tissue developing in the NLG, which are used for detecting chemical information (Brown, 1968), and a reduction in obtaining chemical cues could affect the detection of prey, mates, and competitors.

This led Jaeger (1981b) to test, in E_3, the costs of losing a contest. He partitioned the losers from E_1 into those bitten on the snout ($N = 12$), those bitten on the tail ($N = 6$), and the biting winners ($N = 12$). After 28 days, these salamanders were placed into separate chambers and given 25 prey to eat. He observed the number of prey eaten by each salamander and then examined their NLG microscopically. The 12 snout-bitten salamanders showed scar tissue in the NLG and had significantly lower prey-capture rates than tail-bitten or unbitten salamanders, but the latter two groups did not differ significantly. Adults ($N = 144$) from the forest floor in Virginia were also inspected for scar tissue in the NLG and were sexed, weighed, and measured for SVL. Jaeger found NLG scar tissue in 11.8% of the adults found on the forest floor; those that had NLG scar tissue weighed significantly less than uninjured conspecifics, and both males and females had been involved in biting aggression. Therefore, losing a contest yielded the high risk of long-term injury to the NLG, which can result in a decline in the ability to find prey and potentially cause lower future fitness. He inferred that (1) both males and females of *P. cinereus* display dear enemy recognition; (2) threat and submissive behaviors inhibit the escalation of contests to biting; (3) escalation is more likely to occur when a resident has established prior residency by marking an area with pheromones; (4) when bites occur they are usually to areas of the body that can cause the most harm, such as the tail and NLG; and (5) injury to the NLG results in a decrease in foraging success.

3.6 THE EXPULSION OF INTRUDERS

Based on Jaeger and Gergits' (1979) four criteria needed for territoriality to occur in *P. cinereus*, we now have evidence for red-backed salamanders exhibiting site tenacity, advertisement of defended areas with chemical cues (pheromones) and visual displays, and aggressive defense of areas against conspecific intruders. Here we address the fourth criterion: adults of *P. cinereus* can usually expel intruders from their defended areas. Jaeger, Kalvarsky, et al. (1982) tested the hypothesis of territoriality in *P. cinereus* using three predictions (P): P_1: the defender should show less threat and submissive behavior when alone than when with an intruder, P_2: the defender of an area should perform more aggression and less submission than the intruder, and P_3: the defender should be able to exclude the intruder most of the time. Adult salamanders ($N = 68$), collected from Hawksbill Mountain in autumn, were randomly partitioned into pairs. Each salamander was placed into a 3.8-liter clear glass jar that was then placed mouth to mouth with a jar that held a given salamander's pair; the two jars were separated by a partition for one month so that the paired salamanders could not physically, visually, or olfactorily communicate. After one month of separately feeding each pair, the pairs were tested for territorial behavior by removing the partition for 1 hour and recording the behavior of each salamander. Jaeger, Kalvarsky, et al. defined three behavioral states for comparison: (1) residents, where each salamander was on its own marked area in its own jar (control); (2) intruder, the salamander who entered an area of another; and (3) defender, the salamander whose area was invaded. Intrusions occurred in 6 of 13 male pairs, 3 of 9 female pairs, and 5 of 12 male–female pairs. They found no correlation in the results based on salamander size, and male and female salamanders did not significantly differ in their behavior toward other males or females. Therefore, under experimental conditions, salamanders did not alter their agonistic behavior on the basis of sex. Jaeger, Kalvarsky, et al. also found no significant differences in FLAT among residents, intruders, and defenders, but defenders and intruders spent significantly less time in FTR than residents, while defenders and intruders spent significantly more time threatening and chemoinvestigating than residents with territories that were not invaded. Out of 68 salamanders, there were 50 intrusions. When the defender bit the intruder, the intruder left the area significantly more often than the defender. When there were no bites during intrusions, the intruder still left the area significantly more often than the defender. Therefore, the expulsion of intruders was through advertisement of the area, threat displays, and biting. Defenders were two times faster at biting than intruders, suggesting that they were more willing to escalate contests; 74% of defenders won the contest while 18% of intruders won the contest.

Jaeger, Kalvarsky, et al. concluded that *P. cinereus* is territorial (at least in the laboratory), because salamanders can usually expel intruders from advertised and defended areas.

3.7 TESTING TERRITORIALITY IN THE FOREST

Observing territoriality in the laboratory does not necessarily mean that *P. cinereus* is territorial in the forest. In this section we review the papers that confirmed territorial behavior by *P. cinereus* in its natural habitat in Virginia (Fig. 3.2). Mathis (1989) asked whether the feeding areas (presumed territories) of *P. cinereus* were also used during the courting season. She tested two alternative hypotheses:

> H_1: The spatial distributional patterns in *P. cinereus* are the same in the autumn courtship season and summer noncourtship season (the null case).
>
> H_2: The spatial distributional patterns reflect greater clumping of individuals in the courtship season than in the noncourtship season.

Figure 3.2 One of our research sites at Mountain Lake Biological Station, Virginia. Note the abundance of rocks and logs (cover objects) dispursed in the leaf litter on the forest's floor.

These hypotheses were tested on the forest floor at Mountain Lake Biological Station (MLBS), Virginia, in three ways. Mathis (1989) compared (1) the number of cohabiting individuals found under cover objects in the summer with the number observed in the autumn, (2) summer and autumn interindividual distances for individual salamanders found under the same cover objects, and (3) whether cover objects with multiple individuals were large enough to contain separate home areas for each individual in both the summer and autumn.

On the forest floor, Mathis (1989) established two areas (100 × 200 m), one for summer and one for autumn, where she recorded the number of individuals under cover objects. To determine if the two areas were similar, she placed 10 plots, each 5 × 5 m, within each area and recorded the number and size of potential cover objects ≥100 cm^2 and the number and types of trees and shrubs. A 1 × 1 m plot in the corner of the 5 × 5 m plot was used to compare the percentage of the plot with leaf litter and herbaceous material. She found that the two areas were similar because they did not significantly differ in the percentage of leaf litter or herbaceous cover, the number of potential cover objects, the size of cover objects, the number of trees, or the number of shrubs. She found no significant differences between seasons in the number of salamanders cohabiting under cover objects or in the distance separating salamanders. She also examined cover object sizes to determine if multiple salamanders were found under cover objects that were large enough to contain separate home ranges (estimated as 388 cm^2) for each individual. Single salamanders used cover objects <388 cm^2 significantly less than expected by random chance during the summer and autumn, and in both seasons cover objects <776 cm^2 were used by two salamanders significantly less often than expected by chance. Therefore, Mathis concluded that the data were consistent with the hypothesis that individuals are territorial on the forest floor and that these territories are used during the courting and noncourting seasons.

Using a removal experiment at MLBS, Mathis (1990a) examined the effect of resource quality and body size on territoriality in *P. cinereus*. She tested five hypotheses:

H_1: The removal of a resident from a cover object allows the cover object to be invaded by a new salamander.

H_2: Invading individuals should be smaller than original residents, as owners should be larger than floater individuals that do not hold territories.

H_3: In summer, larger cover objects are cooler than smaller cover objects or leaf litter and, thus, larger cover objects are higher in quality.

H_4: Individuals should prefer larger cover objects, and larger cover objects are less abundant than smaller cover objects.

H_5: Individuals compete for cover objects, and larger salamanders are better competitors.

Figure 3.3 A plot of the forest at Mountain Lake Biological Station near where Mathis (1990a) established her censusing plots.

In experiment 1, Mathis (1990a) conducted a removal experiment on a 6500 m^2 plot where 40 individuals found under natural cover objects (Fig. 3.3) were randomly assigned either to controls or to the experimental group. Control salamanders were removed from under cover objects, uniquely toe-clipped for identification, and then replaced under their cover objects. Salamanders in the experimental group were permanently removed from their cover objects. The cover objects were not disturbed for 5 days. Thereafter, Mathis checked the cover objects for salamanders every 2 days until 14 days had elapsed. She found that under experimental cover objects, where residents had been removed, invasion by new salamanders occurred significantly more often than under control cover objects, where residents had not been removed. Also, the original residents were significantly larger (in SVL) than newly invading salamanders.

In experiment 2, Mathis (1990a) examined cover object choice in the laboratory. Salamanders from MLBS (N = 25 large salamanders, ≥ 42 mm SVL, and N = 26 small salamanders, ≤ 38 mm SVL) were placed into testing chambers (13 × 13 × 8.5 cm) that contained two, randomly positioned plastic tubes (artificial cover objects): one tube was large (20 cm long) and the other was small (10 cm long). She introduced one salamander into each chamber for 24 hours and then noted the location and tube choice of each salamander. If the salamander was not in a tube, then the chamber was rechecked every 4 hours. After five days, the tubes were examined for fecal pellets. She found that both large and small salamanders preferred the larger cover objects and deposited their fecal pellets under these larger cover objects.

In experiment 3, Mathis (1990a) examined cover object choice in the forest at MLBS. She placed 16 large (23 × 24 × 2 cm) and 16 small (11 × 11 × 2 cm) wooden boards on the soil surface and weeks later lifted the artificial cover objects and recorded the presence of salamanders. She found that larger cover objects in the forest housed significantly more salamanders than smaller cover objects.

In experiment 4, Mathis (1990a) examined the distribution of salamanders in their natural habitat. She censused a 260 × 135 m area of the forest floor by turning over all natural cover objects (i.e., measuring ≥ 25 cm^2) and captured the salamanders present so that they could be measured for SVL and sexed (adults only, ≥ 32 mm) during the summer noncourtship season. In a second area, adjacent to the first, she censused during the autumn courtship season. She found that salamander size was significantly and positively correlated with cover object size in the summer and autumn for all salamanders, both female and male adults.

In experiment 5, Mathis (1990a) examined the soil temperature under large and small cover objects. Her hypothesis was that during warmer periods, the substrate temperature beneath larger cover objects should be lower than beneath smaller cover objects or the leaf litter. She measured the soil and air temperature under the center of the cover objects. She found that soil temperatures beneath large cover objects were significantly lower than beneath small cover objects or the leaf litter, while the soil temperature under small cover objects did not significantly differ from the soil temperature in the leaf litter.

The five experiments performed by Mathis (1990a) were important because they were the first set of experiments to provide evidence that cover objects in a natural habitat (MLBS) are defended by individuals of *P. cinereus*. So from these data, we could infer that individuals of *P. cinereus* are territorial in the dry conditions of Virginia, at least at our research site in the southwestern part of the state and in the laboratory, because they defend cover objects against intruders. Mathis also concluded that larger salamanders are superior competitors to smaller ones in contests over resources, and individuals compete for high-quality resources (larger cover objects). Smaller salamanders tend to be excluded from larger cover objects, and the size of cover objects may be one indicator of territorial quality in a natural population and thus important in determining the outcome of a territorial contest. By contrast, in a separate study, Quinn and Graves (1999) found that individuals of *P. cinereus* in northern Michigan, where the forest floor is nearly always damp, are not territorial.

Next, Mathis (1991b) conducted a mark–recapture study at MLBS to determine the home ranges of 107 red-backed salamanders in the natural habitat and to determine the impact of body condition (using tail length as the measure of condition) on territorial and nonterritorial salamanders. She hypothesized that:

> H_1: There is a difference in relative tail length between territorial and nonterritorial salamanders because (1) if body condition is important to the outcome of a territorial contest, then territorial owners (i.e., successful competitors) should be in better condition (have longer tails) than their nonterritorial counterparts and (2) if territorial ownership confers energetic benefits, then territorial salamanders should have more energy reserves stored in their tails (i.e., have longer tails) than nonterritorial salamanders.
>
> H_2: There is a differential cost (that can be measured as tail loss: i.e., by tail autotomy) between territorial and nonterritorial salamanders.

She gave four reasons why territories may be important: (1) there is a relationship between body size and size of home range as larger salamanders have larger energetic requirements; (2) home areas of *P. cinereus* are more segregated than would be predicted by chance; (3) there is a nonrandom association of home areas occupied by males and females so that home areas should be positively associated intersexually and negatively associated intrasexually; and (4) there is a greater intersexual overlap of home areas than intrasexual overlap, because if mating occurs between individuals whose territories overlap, then adults may be more tolerant of intruders of the opposite sex than intruders of the same sex.

Mathis (1991b) placed five 3 × 3 m plots in about 2600 m^2 of forest with marked stakes at 1 m intervals, which she censused 51 nights for surface activity of *P. cinereus*. She located salamanders in each plot, collected them, and placed them into vials left at the point of capture. Once all of the salamanders had been collected on a plot, she removed each salamander (in the order of capture), toe-clipped it, recorded any other distinguishing features, recorded tail condition, measured SVL and total length (TL), and sexed the adults before releasing each individual at the point of capture. She also recorded the location of each individual within the plot on a gridded data sheet. Because nonterritorial salamanders were less likely to be recaptured than territorial individuals, she used the number of recaptures to approximate which salamanders were territorial and which were not; thus individuals that were captured several times were treated as territorial.

Mathis (1991b) captured 107 individuals, 26% of which were juveniles; 48% of the 79 adults were males and 52% were females. Fifty-one percent of the salamanders were recaptured at least once: thus they were considered to be territorial. The density of salamanders on the plots was 2.82 salamanders/m^2. She found that the SVL of salamanders with multiple recaptures was greater than those captured only once. Salamanders presumed to be territorial had significantly longer relative tail lengths (tail = 43% of TL) than presumed floaters (tail = 39.3% TL). Contrary to Mathis's prediction, territorial salamanders had significantly more evidence of tail autotomy than floater individuals, perhaps from defending their territories against intruders (see Jaeger, 1981b, in section 3.5, for biting of tails during aggressive conflicts). There was no significant difference in the size of home areas occupied by adult males, adult females, and juveniles. SVL was significantly negatively correlated with the size of the home area. For both adult males and females, home area centers were significantly further apart intrasexually than would be expected from random distributions, but when all the adults were considered in the analysis, the observed distribution did not differ significantly from the expected random distribution. One inference from this result was that perhaps intersexual areas were closer together than expected by chance. Using a different analysis, Mathis found that the nearest neighbor for 68% of the females was a male, while the nearest neighbor for 73% of the males was a female. Therefore, home areas of *P. cinereus* in their natural habitat are distributed nonrandomly with respect to sex of neighboring individuals, because both males and females are segregated intrasexually but positively associated intersexually, suggesting that residents may be more tolerant toward intruders of the opposite sex than of the same sex. Mathis inferred that some individuals of *P. cinereus* are territorial while part of the population is not (floaters), that these territorial individuals tend to be larger than floater individuals, and that territorial individuals experience more tail autotomy than floaters. Thus we can now infer that *P. cinereus* is not only territorial but that these territories are important feeding areas during dry weather and perhaps play a role in mating.

Toll et al. (2000) performed additional experiments where adults were removed from cover objects near MLBS to test the null hypothesis of no significant difference in certain morphological and behavioral traits between males and females found under cover objects or between male and females invading those cover objects after the residents were removed. Residents were defined as salamanders defending territories (cover objects), and invaders were those salamanders entering territories after the residents had been removed. More specifically, they tested these hypotheses:

H_1: That male and female residents do not differ significantly in SVL, TL, size of cover object inhabited, or number of individuals homing after displacement.

H_2: That male and female invaders do not significantly differ in SVL, TL, size of cover object invaded, or numbers invading uninhabited cover objects.

H_3: That there would be no significant difference between the sex of residents and the subsequent invaders to particular cover objects.

Toll et al. (2000) laid out 10 parallel transects that were 10 to 15 m wide and spaced 10 m from adjacent ones. The length of each transect varied from 50 to 175 m long because of streams. They searched one or two transects from 8 AM to 11 AM on days after "measurable" rainfall (eight different days in July) and removed single adult males and single adult females without eggs from cover objects that were >5 m apart. Each resident salamander was sexed, measured for SVL and TL, and toe-clipped to give each individual a unique mark. Also, a uniquely numbered flag was placed at the location of each resident's cover object. At 11:15 AM, they released the residents, in sequence, perpendicular to the transect and 5 m from cover objects while alternating left and right down the transect. They checked the cover objects for salamanders three times during the day of displacement and then for the next three days. They also recorded the time elapsed before displaced salamanders returned to cover objects; time of first invasion (if any) by other salamanders (not the original); and SVL, TL, and sex of each invader.

Toll et al. (2000) displaced 32 female and 53 male residents from cover objects. They found no significant intersexual differences in SVL, TL, or size of cover objects inhabited by residents. Five males and 13 females invaded the vacant cover objects. These invaders did not significantly differ in SVL, TL, or size of cover objects invaded. Significantly more females than males invaded compared to the ratio of females and males originally displaced, and those females invaded in less than half the time of males. Also, there was no significant difference in whether males or females invaded the territories previously "owned" by male or female residents. Because more females invaded unoccupied territories and did so in less time than males, this suggests that intersexual differences exist for salamanders searching for unoccupied cover objects. These results led to three a posteriori hypotheses:

H_1: Males dominate territorial spaces, with large numbers of females as "nonterritorial floaters" during the noncourtship summer.

H_2: Males hold higher quality feeding territories than do females, so once males are removed a neighboring female can leave her territory in favor of a vacated one.

H_3: Females are more willing to risk predation than males because they need to maximize foraging rates as yolking eggs and guarding clutches are costly (Ng & Wilbur, 1995).

3.8 VARIABLES THAT AFFECT TERRITORIAL CONTESTS

Having established that *P. cinereus* is territorial both in the laboratory and on the forest floor and having defined its agonistic behaviors, we will now describe the numerous variables that influence territorial defense.

3.8.1 Length of residency

Nunes and Jaeger (1989) tested the hypothesis that long-term territorial residents are more tenacious in territorial defense than are short-term residents. This hypothesis originated from Krebs (1982), who suggested that territorial resources (for the great tit: *Parus major*) may be more valuable to an owner than to an intruder because of the owner's superior familiarity with the quality or quantity of resources in the area. Nunes and Jaeger predicted that the longer a male salamander occupies a territory, the more aggressive and the less submissive he should be toward an intruder. They collected adult male salamanders from MLBS and tested them in the laboratory at The University of Louisiana at Lafayette (ULL). Randomly chosen resident male salamanders ($N = 26$) were placed into experimental chambers (29.5 × 22.0 × 4.7 cm) that contained 3 cm of damp soil and an opaque glass vial (length 9.5 cm, diameter 2.4 cm), randomly placed at one end of the chamber in a horizontal position that was used as a burrow. Resident salamanders were allowed either 4 days or 12 days to establish a territory. During this time, future intruders were also housed in separated experimental chambers. Afterward, both the resident and the intruder were placed under opaque plastic covers in the resident's chamber for a 15-minute habituation period prior to a 30-minute test period. They recorded the time inside the burrow, in ATR, FLAT, EDGE (they called it Escape), and the number of NT by both residents and intruders. Each resident was tested in the following four treatments (in random order) with one week between each treatment: (1) 4-day resident encounters an intruder, (2) 12-day resident encounters an intruder, (3) 4-day resident does not encounter an intruder (control), and (4) 12-day resident does not encounter an intruder (control). They found that 12-day residents spent significantly more time in ATR than 4-day residents, 12-day residents were more aggressive toward intruders than 4-day residents, both 4-day and 12-day residents spent more time in ATR toward intruders than toward controls, 12-day residents spent significantly less time in FLAT (less submission) than 4-day residents, only 12-day residents took longer than controls to enter burrows for the first time, 12-day residents exhibited more NT than their controls, and 12-day residents bit intruders approximately three times more often than 4-day residents. Four-day residents spent significantly more time in EDGE than 12-day residents or 4-day residents in the control treatments, indicating that 4-day residents were more likely to abandon their territories to intruders. Intruder behavior did not differ significantly when encountering 4-day or 12-day residents. Overall, 12-day residents exhibited significantly more threat displays and less submissive displays and escape behavior toward intruders than did 4-day residents. Therefore Nunes and Jaeger inferred that increasing knowledge of a territory, through an increased length of ownership, leads to more tenacious defense; thus long-term residents are less likely to be ousted from a territory by an intruder.

3.8.2 Body size

Mathis (1990a), discussed in section 3.7, proposed that larger individuals of *P. cinereus* exclude smaller conspecifics from cover objects through aggressive behavior, and Jaeger, Kalvarsky, et al. (1982, in section 3.6) found that residents expel intruders of equal or smaller body size, but what happens between salamanders with conflicting asymmetries (*sensu* Maynard Smith & Parker, 1976)? Townsend et al. (1998) assessed the ability of small adult males of *P. cinereus* to defend "territories" (constructed in the laboratory) against intrusions by larger male conspecifics. They tested three alternative hypotheses:

H_1: A smaller salamander with resident advantage can repel an intruder of a larger body size.

H_2: An intruder of a larger body size can displace a smaller individual with resident advantage

H_3: The null case.

They predicted that the ability of a small resident to defend a territory successfully would decrease as the magnitude of the size asymmetry between resident and intruder increased.

Townsend et al. (1998) collected 120 adult males from MLBS, housed them in Petri dishes with potting soil, and fed them termites (*Reticulitermes*). Prior to testing, the salamanders were partitioned into four size classes ($N = 30$ each): 34–36 mm SVL, 37–39 mm SVL, 40–42 mm SVL, and 43–46 mm SVL. The salamanders in the smallest size class were residents that were tested with a randomly chosen salamander from one of the other three size groups (without replacement), and once with no intruder (control), in random order. Thus the size differences between residents and intruders increased from 3 mm, to 6 mm, to 9–10 mm. On day 1, residents were introduced into laboratory constructed territories that consisted of two large Petri dishes (150 × 15 mm) joined together by a small plastic tube (75 × 15 mm). Researchers introduced the residents into the territory by a hole in the tube that was then plugged. Each Petri dish was covered by black construction paper to act as a hiding place, and one of the Petri dishes (chosen at random) contained a small board with termites on it at the start. The chambers were left alone for 5 days. On day 6, Townsend et al. introduced an intruder salamander from one of the size classes to the resident via the central tube. Every 30 minutes for 5 hours they observed the positions of the resident and intruder in relation to the position of the initial food-rich side. A win was recorded if the resident was in the initial food-rich Petri dish, a loss was recorded if occupancy was by the intruder, and a draw was recorded if the salamanders shared occupancy. They found that residents significantly preferred to occupy the side containing food when tested alone. Residents also achieved exclusive occupancy of the side containing food significantly more often than did intruders. Therefore, having a favorable asymmetry, territorial residency, can partially offset an unfavorable asymmetry in body size (smaller body size) in agonistic encounters between adult males of *P. cinereus*. Smaller residents were more successful at expelling the larger intruders than the larger intruders were at expelling the smaller residents. Thus a territorial owner (resident) may have more to gain by winning (keeping the territory) or more to lose if defeated (losing the territory; Maynard Smith & Parker, 1976) than an intruder trying to gain a new territory.

Gabor (1995) also examined the impact of body size on the ability to gain or maintain residency of high-quality territories. She tested Mathis's (1990a, in section 3.7) hypothesis that bigger salamanders have better territories. She correlated quantities and profitability of prey available within territories of *P. cinereus* to body size and condition of salamanders holding those territories. She specifically tested the hypothesis that larger salamanders, based on SVL and mass, would be found in more profitable food territories, based on number of prey and prey type, than would smaller salamanders. She assayed the invertebrate population around the perimeter of 41 territories held by adults (under rocks and logs >50 cm^2) of *P. cinereus* at Hawksbill Gap, SNP, during a 4-day period in July 1992 on two 2500 m^2 sites 9 days after 1.96 cm of rainfall. At each site, she placed 7.6 × 12.7 cm (bright yellow plastic sheets) sticky traps for invertebrates up to the size of large cockroaches. One trap was placed at each of the four coordinate directions at the litter–soil interphase approximately equidistant around the territory. Each trap was protected by a 14.5 cm Petri dish that was placed approximately 3 cm above the trap, supported by sticks pushed into the ground and covered with leaves. She collected each trap 24 hours after being set and placed it into a polyethylene bag, which was frozen until examination. She also recorded the greatest width and length for each cover object, counted the number of invertebrates in each trap, identified the prey to taxonomic order, measured TL and total width of each prey item, and determined the volume of each prey following Jaeger's (1980b) procedure. Gabor only counted prey that *P. cinereus* eats and ranked these prey as high if soft bodied and as low if hard bodied, based on criteria from Jaeger (1990, see section 4.2) and Gabor and Jaeger (1995). She measured SVL and TL of each of the 41 territorial salamanders, determined its sex, and estimated its physical condition.

Gabor (1995) found that the number of prey around each territory was significantly positively correlated with salamander SVL and mass. There was a significant positive correlation between the mass as a function of SVL of salamanders, and a significant, positive correlation between the number of Diptera (soft prey) and salamander SVL and mass. Neither total volume of prey per territory nor number of Acarai (hard prey) were significantly correlated with salamander SVL or mass. There were no significant correlations between the number of Diptera or Acari on traps and cover object size, or between the size of salamander cover objects and salamander SVL or mass. Therefore, the data supported Mathis's (1990a) hypothesis that bigger salamanders have better territories, but Gabor's data did not support Mathis's idea that larger cover objects may allow salamanders access to a larger number of prey items. Gabor inferred that either bigger salamanders gain access to food-rich territories or salamanders that happen to be in food-rich territories grow bigger than those in poorer quality territories. Thus territories serve as feeding areas.

3.8.3 Body size without residency

Townsend and Jaeger (1998) tested the hypothesis that possession of a relatively large body size is advantageous in agonistic encounters for food between adult males of *P. cinereus*. They utilized a resident–resident design that allowed them not only to minimize the effects of other potentially conflicting asymmetries (e.g., prior residency) but also to examine and describe the effects of relative body size on the expression of aggressive behavior in agonistic interactions in two different size classes of salamanders. Sixty adult males were collected from MLBS and tested in the laboratory at ULL. The salamanders were divided into two size classes ($N = 30$ each): large salamanders measured 44–49 mm SVL and small salamanders measured 34–39 mm SVL. The test chambers ($22 \times 12 \times 3$ cm) were lined with a paper towel, divided into three sections by two removable aluminum dividers reinforced with duct tape, and covered with glass. Townsend and Jaeger tested each salamander three times (once in each of the three treatments for its size class) in random order. These treatments were (1) one large individual confined in each end of the test chamber, (2) a large individual housed with a small individual so that the size difference was 6 to 8 mm SVL, (3) a large individual housed with a surrogate control (rolled piece of paper towel the size of a salamander), (4) two small individuals housed together, and (5) a small individual housed with a surrogate. In treatments 1 and 4, the salamanders were size-matched to 1 mm SVL. On day 1, researchers placed one or both salamanders, depending on the treatment, into the end sections. On day 2, they placed six termites (*Reticulitermes*) into the central unoccupied area, randomly removed one of the barriers, and allowed the salamander to feed for 30 minutes. After 30 minutes, the barrier was restored and the salamander was placed back into the end chamber. The paper towel in the center chamber was changed, six more termites were added, and the second barrier was removed so that the second salamander could feed for 30 minutes in conditions 1, 2, and 4. After 30 minutes, the barrier was replaced and the salamander was returned to the end chamber. On day 4, the same procedure occurred as on day 2. On day 6, researchers placed 20 termites into the central area, both barriers were removed at the same time, and they observed each salamander for 15 minutes. They recorded the number of termites eaten and the total time in ATR by each salamander.

When pairing larger versus smaller salamanders, larger salamanders spent significantly more time in ATR and ingested significantly more prey than did smaller salamanders. Larger salamanders did not exhibit a significant difference in time in ATR or in the number of termites eaten when paired with smaller individuals as compared to surrogates or similar-sized individuals. Smaller salamanders did exhibit significant differences in ATR. They spent significantly more time in ATR when paired with similar-sized individuals as compared to surrogates. Smaller salamanders also ingested significantly fewer termites when paired with larger conspecifics than when paired with surrogates. Townsend and Jaeger (1998) inferred that relative body size is a significant factor in determining the outcome of contests for prey between territorial males of *P. cinereus*, because larger males displayed more aggressive behavior and ingested more prey than did smaller males. Also, smaller salamanders were more likely to alter their foraging and aggressive behavior in response to conspecifics compared to larger salamanders, which were not affected by size or the presence of an opponent. Thus a body size asymmetry matters when territorial ownership is eliminated from a contest.

3.8.4 Sex and reproductive condition

Horne (1988) compared the territorial agonistic behavior of known nongravid (noncourting) females ($N = 26$) to that of gravid (courting) females ($N = 26$) of

P. cinereus. She tested the hypothesis that gravid female residents will defend territories against all conspecific intruders except males. Adult salamanders were collected in Virginia and tested in the laboratory at ULL during spring. The test chambers (40.2 × 16.8 × 1.3 cm) contained opaque plastic burrows (7.5 × 2.0 × 0.8 cm), covered with damp paper towels, which were placed at one end of each chamber. The salamanders were separated into two randomly chosen groups: residents (gravid and nongravid females) who, after 6 days of residency remained in their own chambers during the test, and intruders (males, gravid females, and nongravid females), who were allowed to establish territories in the same way as residents and were introduced into the residents' chambers on day 6. Horne introduced the intruder into the resident's chamber by taking it from its own chamber and placing it under a habituation cup (and the resident under a separate cup) for 15 minutes prior to recording the behavior of both salamanders. She paired residents and intruders in six combinations and then compared (1) gravid and nongravid residents, (2) intruders in territories of gravid and nongravid females, and (3) residents and intruders.

Horne (1988) found that gravid female residents spent significantly more time in FLAT than nongravid female intruders when faced with nongravid female intruders. Nongravid female residents showed significantly more FLAT behavior toward male intruders than gravid females, and nongravid female residents displayed significantly more EDGE when the intruders were gravid females or males than when intruders were males compared to gravid residents. Male intruders spent significantly more time in EDGE in the nongravid female's chamber than in the gravid female's chamber and significantly more time in the burrow when invading territories of gravid females than when introduced to territories of nongravid females. Gravid female intruders spent significantly more time in EDGE than gravid female residents and nongravid female residents, and male intruders spent significantly more time in FLAT than gravid female residents. Horne inferred that gravid females were as aggressive toward intruders as were nongravid females but differed in submissive behavior toward intruders. Therefore, she proposed that gravid females continue to inhabit their territories during the breeding season and that gravid females are more likely to share their burrows with males than nongravid females.

3.8.5 Intruder number

Kohn et al. (2005) examined the possibility that the cost of territoriality increases when the number of intruders simultaneously entering an adult female's territory increases from zero, to one, to two. They tested the null hypothesis that time in ATR, EDGE, and number of bites by *P. cinereus* would not differ significantly among the five (randomly tested) treatments: (1) one female intruder, (2) two female intruders, (3) one male intruder, (4) two male intruders, and (5) a surrogate control (rolled piece of paper towel). Alternatively, female aggression could increase as the number of intruders simultaneously entering her territory increased, or the potential cost of keeping her territory could become too great with the increased number of intrusions so that she might decrease her aggressive behavior and increase her escape behavior. Single adult salamanders (130 females and 106 males) were collected at MLBS and tested at ULL. The salamanders were size matched (within 4 mm of each other), and one of four females was randomly chosen as the territorial resident that then encountered one or two female or male intruders. Each resident female ($N = 30$) went through all five treatments in random sequence. On day 1, Kohn et al. placed each focal female into a Nunc bioassay chamber (24 × 24 × 2 cm; Fig. 3.4) that contained a moist paper towel and fed the salamanders eight *Drosophila virilis*. On day 4, the salamanders were again fed eight fruit flies and the paper towel was moistened with spring water. On day 5, researchers introduced one female, two females, one male, two males, or a surrogate into the female resident's chamber and then recorded her behavior for 15 minutes.

Kohn et al. (2005) found a significant difference in behavior across the treatments for ATR, EDGE, and bites. Post hoc comparisons found that, relative

Figure 3.4 A female red-backed salamander in a Nunc bioassay dish.

to the control, females displayed significantly more ATR toward one male intruder and significantly more EDGE in the presence of two male intruders. There was no significant difference toward one or two female intruders. They inferred that, for females, the benefit of guarding prey in a territory outweighed the cost of defense (possible injury) when a single male was in her territory, but the costs exceeded the benefits when two male intruders were simultaneously introduced to her territory. Therefore, the cost–benefit ratio of territorial defense for female *P. cinereus* may depend on the sex of the intruder, the number of simultaneous male intruders, and perhaps whether the females do or do not cohabit a territory with a male partner (see section 7.10 for more on male–female partners).

3.8.6 Tail condition

Thus far we have found that prior residency, length of residency, body size, sex, and intruder number impact territorial defense. Now we turn to the influence of tail loss resulting from tail autotomy. Wise and Jaeger (1998) studied the independent and collective effects of residency status (resident/intruder) and tail condition (tailed/tailless) in staged encounters between adults of *P. cinereus* (that initially had tails) from MLBS in the laboratory at ULL. They predicted that residents and intruders would alter their behavior based on their own tail condition and the tail condition of their opponents. They also predicted that individuals would be more aggressive (and/or less submissive) when there were small, but not large, tail asymmetries between opponents.

Groups of four individuals (tetrads) were size matched (within 1 mm SVL), and within each tetrad (N = 28) Wise and Jaeger (1998) randomly chose the future tailed resident, tailless resident, tailed intruder, or tailless intruder. They randomly tested all combinations of resident–intruder pairs with one week between tests. They induced tail autotomy prior to the start of the experiment by grasping a salamander's tail with forceps and allowing the salamander to autotomize its tail. Tailless salamanders had 85% of the tail removed while tailed salamanders had 2 mm of the tail removed as a sham control. On day 1, Wise and Jaeger placed residents and intruders in test chambers (31 × 17 × 1.5 cm) lined with two layers of moist paper towel, and the salamanders were fed *Drosophila virilis ad libitum*. On day 5, they introduced the intruder salamander into the resident's chamber, covered both salamanders with a transparent cover object for a 15-minute habituation period, and then observed the behavior of both salamanders for 30 minutes. They also remeasured tails 15 days following the experiment as the tails partially regenerated, but length of tail regeneration did not significantly impact the experimental results.

Wise and Jaeger (1998) found that neither residents nor intruders altered their behavior based on their own tail condition, but they did alter their behavior based on the tail condition of their opponents. Intruders showed significantly more aggression or less submission toward tailless residents than toward tailed residents. When contests were between residents and intruders of the same tail condition (both tailed or both tailless), intruders were more aggressive toward residents when both were tailless than when both were tailed. Thus tail loss did not directly hamper aggressive displays. When the asymmetry between residents and intruders was small, intruders showed more aggression and less submission than in contests where the asymmetry was large. Therefore, tail loss influences agonistic behavior and thus may be an important determinant of success in territorial contests. For residents and intruders, the tail condition of their opponents was more important in influencing decisions concerning territorial behavior than was their own tail condition. Wise and Jaeger also inferred that intruders assess the physical condition of residents and are more aggressive when there is a small asymmetry between contestants because there is a higher probability of winning the contest.

3.8.7 Food quantity

Another variable that impacts territoriality is based on food resources. Nunes (1988) studied the interactions between males of *P. cinereus* when both contestants were residents of neighboring territories but there was an asymmetry in the amount of food received by each individual ("low" or "high" food levels). She tested the alternative hypotheses:

> H_1: Salamanders maintained on low food levels prior to the contest would be more aggressive because they should be "hungrier" and thus need more food.
>
> H_2: Salamanders maintained on high food levels should be more aggressive because the value of a high-quality food territory should exceed the cost of being injured in a fight.
>
> H_3: Salamanders fed low food levels should not differ in aggressiveness from salamanders fed high food levels during territorial disputes (the null case).

Male salamanders were collected from MLBS and tested in the laboratory at ULL. The experimental chambers (29.5 × 22.0 × 4.7 cm) contained 3 cm of soil with three layers of damp paper towels, a wooden partition that divided the chamber in half and ran across the width of the chamber, and an opaque plastic tube (length 9.5 cm, diameter 2.5 cm) containing a sponge in the middle that divided the tube in half and was used as a burrow. The plastic tube was pushed through a hole in the wooden partition. Four days prior to the test, a randomly selected pair of size-matched males ($N = 25$) were placed into the chamber so that one male was on each side of the wooden partition. The salamanders were allowed to establish territories for 4 days and were fed daily with *Drosophila virilis* that had been freeze-killed and then thawed just before feeding. One of the salamanders was randomly selected to be given a low food level (~3–6 flies per day as calculated by Jaeger, 1980b, in section 3.2) so that the salamander did not gain or lose body mass. The other salamander was maintained on a high food diet (i.e., double the amount of the low food diet). Ten minutes before testing, any remaining food was removed. Just before testing, the wooden partition was removed and a plastic tube was placed into the center of the chamber where the original burrow had been located so the salamander could move freely between the sides. For 30 minutes, Nunes recorded the time spent in the burrow, in the opponent's area, and in ATR, FLAT, EDGE, as well as the number of NT to the substrate or to fecal pellets. Pairs of salamanders went through the following four treatments, in random order, with one week between treatments: (1) one individual was maintained on low food levels while the other was maintained on high food levels, (2) the feeding regime was reversed for the pair, (3) both salamanders were maintained on low food levels, and (4) only one randomly selected individual of the pair was placed in the chamber and maintained on low food levels (the control).

Nunes (1988) found that salamanders maintained on low food levels spent significantly more time inside the burrow than when they were maintained on high food levels, NT fecal pellets significantly more in all conditions compared to the control treatments; thus the presence of an opponent elicited chemoreception. Salamanders fed low food levels spent significantly more time NT, NT to fecal pellets, and in EDGE. There were no significant differences in time in the opponent's area or in ATR. Thus salamanders fed low food levels avoided interactions with opponents. These results suggest that the amount of food in a territory affects the behavior of salamanders during territorial disputes.

3.8.8 Food quality

In addition to food quantity, food quality may also impact territorial behavior. Gabor and Jaeger (1995) tested whether the agonistic behavior exhibited by male salamanders was affected by the quality (rather than quantity) of food available in a territory using laboratory experiments. They predicted that prey quality is in part a function of digestion efficiency and gut-passage time for salamanders. They used termites (*Reticulitermes*) as high-quality food and ants (*Solenopsis molestra*) as low-quality food. They tested two hypotheses:

> H_1: Salamanders should pass termites through their digestive tracts in significantly less time than ants and should have a significantly higher digestion efficiency when eating termites than when eating ants.
>
> H_2: Territorial residents and intruders should show significantly more aggressive behavior when the residents have access to a higher quality food resource relative to one of lower quality.

They determined gut-passage time and digestion efficiency by randomly selecting 40 males that were fed ants or termites. No food was given to those males for two weeks prior to testing. On day 1, Gabor and Jaeger (1995) placed 10 termites or 20 ants (ants were half the size of termites) into each salamander's Petri dish (14.5 × 1.5 cm) lined with a moist filter paper. After 1 hour, they counted the number of insects ingested, removed any uneaten prey, and changed the filter paper in the Petri dishes. Of the 40 males, 23 ate termites and 11 ate ants. On day 3, they started recording gut-passage time (feces released) in 0.5 hour increments. The feces were placed in preweighed glass test tubes, wrapped securely with parafilm wax, and stored in a freezer. Later the researchers calculated the dry mass of the feces to estimate digestion efficiency.

To test for differences in territorial aggression, Gabor and Jaeger (1995) tested residents in six different treatments: (1) a territorial resident previously fed termites with an intruder, (2) a resident fed termites with a surrogate (rolled paper towel), (3) a resident fed ants with an intruder, (4) a resident fed ants with a surrogate, (5) a resident fed half the amount of ants and termites with an intruder, and (6) a resident fed half the amount of ants and termites with a surrogate. They randomly chose residents and intruders for each treatment. On day 1, future residents and intruders were placed into separate chambers (31.5 × 17 × 1.4 cm) lined with moist paper towels. Residents were fed 12 to 14 termites, 24 to 26 ants, or 6 to 7 termites plus 12 to 13 ants; intruders were always fed 16 to 18 flies (*Drosophila virilis*). The salamanders were fed on day 1, between days 3 and 4, and in the morning of day 6. On day 6, researchers removed any uneaten food and placed the intruder into the resident's test chamber. Both the resident and the intruder were placed under cover objects on opposite sides of the chamber for a 15-minute habituation period prior to a 30-minute test period.

Gabor and Jaeger (1995) found that ants took significantly longer (mean = 152.2 ± 57.4 hours) to pass through the gut than termites (mean = 97.2 ± 28.4 hours). Males also digested significantly more of the termites than of the ants, based on dry mass of the feces. Because termites passed through the digestive tract faster than ants and salamanders digested termites more efficiently than ants, Gabor and Jaeger inferred that termites are a higher quality prey than ants.

In terms of aggression, residents fed termites spent significantly more time in ATR than residents fed ants with intruders present. There was no significant difference between residents fed termites or residents fed ants and termites, but residents fed ants and termites spent significantly more time in ATR than those fed ants. There was no significant difference in ATR toward surrogates when residents were fed termites, ants, or ants plus termites. There was a significant difference in bites toward intruders when residents were fed ants, or termites, or ants plus termites. Residents fed termites bit intruders significantly more frequently than residents fed ants. The other comparisons were not significantly different. Regarding intruders, overall there was a significant difference in ATR by intruders toward residents fed ants, termites, or both. There was significantly more ATR from intruders encountering residents fed termites than residents fed ants. The other paired comparisons were not significantly different, nor was there a significant difference in the number of bites by intruders toward residents fed ants, termites, or both.

Based on these results, Gabor and Jaeger (1995) concluded that residents and intruders show more aggressive behavior when a resident has access to a high-quality resource (termites) than a low-quality resource (ants). They also inferred that intruders were informed about the feeding asymmetry by either (1) directly detecting the chemical cues of termites or intruders or (2) changing their response based on the residents' increased aggression. They concluded that intruders were more aggressive in an attempt to gain a high-quality territory. Thus the quality of food resources in a territory affects the agonistic behavior of residents (and possibly intruders), a hypothesis tested next.

3.8.9 Signal honesty

Gabor and Jaeger (1999) tested the hypothesis that resident males of *P. cinereus* exhibit honest agonistic signals toward intruders that depend on the food quality in their territories. They also tested the hypothesis that conventional signals may replace honest signals when the cost of fighting is high, because a rich feeding territory is more worthy of defense than a poor-quality territory. They defined a conventional signal as ATR by a resident that is not subsequently followed by biting and an honest signal as ATR by a resident that is subsequently followed by biting the intruder (*sensu* Dawkins & Guilford, 1991). If an honest signal is used, then they predicted that residents in poor-quality territories should bite less frequently and spend more time in ATR before biting than residents in food-rich territories, because the former should avoid retaliatory bites from intruders. Residents in food-rich territories should risk retaliatory bites by spending more time in ATR before biting, and they should bite more frequently. If a conventional signal is used, they predicted that residents in food-rich territories should spend more time in ATR with no subsequent biting than residents in food-poor territories, and residents in food-rich territories should also spend more time in ATR without subsequent biting than ATR followed by a bite.

Gabor and Jaeger (1999) collected adult males of *P. cinereus* with intact tails from Hawksbill Mountain, SNP, and randomly partitioned them into 28 future residents and 28 future intruders that were sized matched. Each male was tested twice in random order. They allowed each resident 5 days to establish territories in chambers (31.5 × 17 × 1.4 cm) lined with moist paper towels. In one test, residents were fed 12 to 14 termites (high-quality food), and in the other test residents were fed 24 to 26 ants (low-quality food), while intruders were fed flies (*Drosophila*): see the previous discussion for Gabor and Jaeger's (1995) rationale. Each resident encountered different intruders in ant versus termite treatments. On day 6, researchers removed any extra prey and then placed both the intruder and the resident under habituation cups for 15 minutes in the resident's chamber, after which they recorded the behavior of both salamanders for 30 minutes. They separated the data into time in ATR before biting (honest signal) and time in ATR with no biting (conventional signal).

Significantly more residents previously fed termites performed ATR than those in territories previously containing ants, and significantly more residents fed termites bit intruders than those fed ants. Therefore, salamanders with territories that had more profitable prey were more likely to display threat and biting behavior against intruders. There was no significant difference in amount of time in ATR before biting by termite-fed residents compared to ant-fed residents. Residents fed termites spent significantly more time in ATR when no bites followed than ant-fed residents. Gabor and Jaeger (1999) inferred that males are not constrained to give honest signals (ATR) about willingness either to bite or not to bite during territorial defense. Therefore, they hypothesized that male territorial salamanders play a mixed evolutionarily stable strategy game (*sensu* Maynard Smith, 1982) between honest and conventional threat signals.

In sum, territorial defense by *P. cinereus* is influenced by prior residency, length of residency, body size, sex, intruder number, tail condition, food quantity, and food quality.

3.9 LIFE HISTORY TRAITS AND TERRITORIAL CONTESTS

Now that we have evidence that many variables impact territorial defense, we turn our attention to how life history traits influence territorial contests. Wise

(1995) studied variation in the reproductive success of females of *P. cinereus*, territoriality in the natural habitat, and variation in payoff and RHP of territorial ownership. We summarize these three topics in sequence.

3.9.1 Variation in reproductive success among females

To determine reproductive success, Wise (1995) collected 54 females that were brooding eggs at MLBS and transported both females and their clutches (3–11 eggs/clutch: mean = 6.9) to ULL. When the eggs hatched in the laboratory, the neonates were not fed for 40 days until their yolk sacs had been absorbed, at which time Wise measured and weighed each neonate. Neonates were then fed *D. melanogaster* and tubifex *ad libitum* until they were 185 days post-hatch. At 40 days post-hatch, Wise regressed the mass of neonates against the SVL of their mothers. At 185 days, she counted the number of oocytes (seen through the abdominal wall) of each mother and, using regression, compared the number of oocytes and the mother's condition (fat reserves) on clutch size and hatch date. She further examined the effect of the mother's size and condition, clutch size, and hatch date on offspring size and condition and on the number of neonates per clutch that had survived until 185 days post-hatch.

Overall, females showed significantly increasing investment in reproduction with increasing body size. The total mass of a female's offspring was greater for larger than for smaller females, as was number of offspring, the mother's clutch size, and survival of her neonates to day 185. Also, neonates that survived to day 185 (29%) were significantly larger than those that died, as measured at 40 days post-hatch when 100% of the neonates were still alive. The offspring that survived were produced by mothers in poorer condition (less fat reserves) just after brooding (which lasts for at least 48 days, during which females are rarely able to feed). This suggests that females that devoted more metabolic resources to a clutch (e.g., by using more fat reserves) were also more likely to have surviving offspring. There was a positive relationship between the number of new oocytes and body size of females and between the condition of females after the hatching of their clutches and the number of new oocytes.

Wise (1995) concluded that females invest heavily in offspring, but this large investment in current reproduction may explain why most females can lay only one clutch every two years and why females with larger clutches produced fewer new oocytes. She concluded that "investment in current reproduction may reduce the amount of energy available for the next reproductive season" (41). Of course such a long period devoted to brooding while ingesting few prey is in itself a considerable investment in offspring. Recall that Mathis (1990a, in section 3.7) found that territories are usually gained and successfully defended by the largest females (and males) in the population at MLBS, which suggests that territory-holding females are likely to be the most successful reproducers.

3.9.2 Payoffs to and RHP of territorial owners

For *P. cinereus*, the payoffs for successful territorial acquisitions are living under a larger cover object (rock, log) in the forest (Mathis, 1990a), where the soil is moister, is cooler, and may contain more invertebrate prey than smaller cover objects (Gabor, 1995; Mathis, 1990a). However, females gain additional advantages from larger rocks, because (1) that is where they attach and brood their eggs (Wise, 1995) and (2) they may need more food than males in order to yolk those eggs. Yet there should be an optimal (not maximal) size of cover object, because the larger it is (e.g., a long log), the more difficult it should be to defend against territorial intruders; this would be especially true at MLBS with its large nonterritorial population of *P. cinereus* moving about as "floaters" (Gillette, 2003). Variation in RHP also influences the ability of *P. cinereus* to gain and to defend territories. As seen in section 3.8, individuals with longer SVL are more successful in territorial tenacity, as are individuals with intact (not autotomized) tails.

Wise (1995) conducted two experiments in the forest at MLBS to gain more insight concerning the variations in payoffs and RHP on the ability of residents to maintain territories. These were (1) short-term (21 days) and (2) long-term (75 days) mark-recaptures of residents as influenced by payoffs (sizes of their cover objects and sexes of the residents) and RHP (their SVL and tail conditions). In experiment 1, she located 128 adult males and 52 adult females, then recorded the sizes of their cover objects and their sexes, SVL, tail lengths, and weights. She then uniquely marked each individual (and its cover object), induced tail autotomy in 70 males and 25 females, photographed and weighed each one, and then released them under their home cover objects. In experiment 2, the same procedures were used except that 80 adults were tested, which were all subjected to 85% tail autotomy (i.e., posterior to the postcloacal gland. Note that *P. cinereus* will autotomize its own tail near the area where it is lightly pinched. These lost tails will eventually regenerate to near full length.)

When recaptured after 21 days (experiment 1), variation in the residents' sex, cover object area, tail condition, and SVL did not significantly affect their residency under cover objects. However, intruders found on territories of missing, tail residents were significantly larger than intruders on territories of tailless residents. Wise (1995) inferred that smaller intruders may be able to displace tailless residents but only large intruders can oust same size–tailed residents, a nice example of tail condition as a signal of RHP for *P. cinereus*. When recaptured after 75 days (experiment 2 where all males and females had been tail autotomized), resident females had retained their cover objects more often than male residents. This suggests that cover objects do provide higher payoffs for females compared to males. This difference in territorial tenacity between the sexes occurred even though sex had no effect on growth rates (i.e., SVL, mass, and tail regeneration) over 75 days.

Wise (1995) thus inferred that (1) the RHP of residents and payoffs provided by their territories influenced the level of territorial defense by *P. cinereus,* and (2) territorial payoffs for females were greater than for males. Females not only need territories to brood eggs, but they also need to ingest abundant numbers of prey to yolk those eggs and to survive a long period of near starvation during brooding.

3.9.3 Resource acquisition and energy allocation

Wise (1995), along with Henry Wilbur (then director at MLBS), examined the effects of body size and tail loss on resource acquisition and energy allocation by *P. cinereus* at MLBS during the summer of 1992. They made two a priori predictions: (1) if competition occurs between opponents for food or space over long periods of time, then variation in RHP will translate as greater growth (body mass) for winners than for losers; and (2) residents should be found under larger cover objects because they can displace intruders from these favored sites, and intruders' displacements should be based on behavior and condition (RHP) of those residents.

They placed enclosures (144 experimental tubs) into the forest floor (Fig. 3.5) and placed a pair of salamanders into each tub: an established resident and a later intruder. The residents were larger than, equal to, or smaller than the intruders and were either tailed or tailless; intruders were also either tailed or tailless. This created 12 experimental conditions with 12 replicates each. The tubs were 91 × 61 × 21 cm with holes for drainage (from rainfall) and contained soil and leaf litter. One larger and one smaller cover object was placed into each tub. Both sex-paired males and sex-paired nongravid females were measured, photographed, and placed into the tubs with residents (plus prey) preceding intruders by 7 days, so as to establish territories before the intruders arrived. Finally, each tub was destructively searched 80 to 81 days later; the positions of the pairs determined (relative to the cover objects); and the salamanders measured again for SVL, tail length, and body mass. Data were analyzed by analysis of variance and multiple logistic regression.

Wise (1995) found that tailless residents gained more mass than tailed residents regardless of the presence of intruders, but tailless residents gained more

Figure 3.5 Experimental tubs similar to those used by Wise (1995) at Mountain Lake Biological Station.

mass than tailed residents only when intruders were tailless. This suggests that tailless residents still have a competitive advantage (e.g., eat more prey) because of residency status but lose that advantage when intruders are tailed. Thus a salamander would have the greatest competitive advantage (RHP) when it is both a resident and tailed and the intruder is tailless, but the resident invested energy into growth only when the level of competition was highest based on competitive asymmetries in RHP between resident and intruder. Wise proposed that residents with a secure competitive advantage (both residency status and tailed) do not allocate energy to growth but instead invest it in territorial defense. Perhaps the best tactic for a tailless salamander is just to regrow its tail (and RHP) quickly, but Gillette (2003; see section 7.19) showed that males do so more than do females at MLBS and that total or partial tail loss is surprisingly common there. Further, Wise found that residents were more likely to be under the larger of the two cover objects, as predicted earlier, regardless of whether the intruders were nearby. Again, this supports the hypothesis that residency confers a competitive advantage in territorial contests.

Wise (1995) inferred from this experiment that the relative RHP of territorial opponents (resident vs. intruder, relative body size, and tail condition) leads to variance in how individuals of *P. cinereus* will acquire resources, space themselves relative to each other, and allocate energy to growth versus agonistic behavior. She also noted that the evolution of tail autotomy (which is useful against predatory attacks) is associated with a radical and rapid change in the RHP and energy allocation in *P. cinereus*. Further studies on this topic could be fruitful for other biologists studying tail-autotomizing species of salamanders and lizards.

3.10 SEASONAL AND GEOGRAPHIC VARIATION IN TERRITORIAL AGONISTIC BEHAVIOR

Wise and Jaeger (2016) examined both seasonal and geographic variation in territorial, agonistic conflicts by males of *P. cinereus*. They posed alternative hypotheses concerning seasonal variation.

> H_1: Males are more intrasexually aggressive in spring and autumn, which are the breeding seasons.
>
> H_2: Males are more aggressive during the noncourtship summer because warmer weather increases metabolism while decreasing the digestive assimilation efficiency of ingested prey (Bobka et al., 1981, in section 2.2), causing increased competition for prey and thus for territories (Wise, 1995, in section 3.9).

3.10.1 Seasonal variation

Jaeger had casually noticed that populations from Hawksbill Mountain, SNP, and from MLBS seemed to behave differently in aggression in laboratory experiments. The former appeared to reduce time in postural displays and progress faster to biting opponents while the latter tended to extend time in such displays and bite infrequently. For tests of seasonal variation, Wise (1995) collected adult males of *P. cinereus* at MLBS during five periods of time: July and September 1993 and May, August, and October 1994. Wise also collected salamanders at Hawksbill Gap, SNP, during those same five months. Some of the salamanders (not collected) had recently, naturally autotomized tails, which allowed for comparison of occurrence of tailed and tailless salamanders in different seasons. Behavioral tests were performed from 7 PM to 4 AM in the laboratory at MLBS or ULL. Typical (see section 3.8) resident versus intruder territorial tests were conducted but with male pairs asymmetrically matched for RHP (residents with resident advantage and intruders with < 1 mm size advantage). Each test lasted 20 minutes, and researchers recorded for both paired individuals total times in ATR and escape/avoidance behavior (EDGE) and number of bites per contest. Behavioral data were analyzed mostly by multivariate analysis of variance and analysis of variance to detect interaction effects. Each of the 10 conditions (five seasons at two localities) was replicated with 21 to 33 pairs of males (total $N = 249$ behavioral tests).

Salamanders from MLBS and Hawksbill Gap showed significant variation in agonistic behavior among seasons and between the two populations. Both residents and intruders at Hawksbill Gap (where *P. shenandoah* is a congeneric competitor) spent significantly more time in EDGE than at MLBS (where no similar-sized congeners occur). Also within contests where intraspecific bites did occur, *P. cinereus*, at Hawksbill Gap, bit significantly more times per contest than did conspecifics at MLBS. Those at Hawksbill Gap tended to bite more severely (bite, hold, roll in wrestling bouts) than those at MLBS (just nip-bites without wrestling). Wise and Jaeger (2016) suggested that the elevated EDGE and biting behaviors at Hawksbill Gap were influenced by the highly aggressive interspecific behavior of *P. shenandoah* in SNP (see section 6.1 for such interspecific contests).

Residents and intruders spent significantly less time in EDGE during the spring courtship season than during either the summer noncourtship or autumn courtship season at both localities. In contrast, they spent significantly more time in ATR during the summer than during the other seasons. Thus male–male territorial aggression was not a direct consequence of intermale competition for mates but was related apparently to intermale competition for territories (and prey) during warm summer weather. During late autumn, males spent significantly more time in EDGE and less time in ATR (combined) compared to either spring or autumn, perhaps due to their movement into fossorial retreats with approaching winter's freezing weather. *Plethodon cinereus* often forms aggregations in underground cavities.

Across seasons, a significantly greater proportion of males at Hawksbill Gap suffered tail autotomy than did conspecifics at MLBS, again suggesting that intraspecific aggression is more intense at Hawksbill Gap than at MLBS. Two alternative hypotheses were posed by Wise and Jaeger (2016):

H_1: Predator pressure is greater at Hawksbill Gap.

H_2: Females there are more aggressive toward males (see sections 7.10 and 7.12 for such aggression by females).

Tail autotomy was significantly greater in May compared to all other seasons in the forests at both Hawksbill Gap and MLBS. This contrasts with the previous results that ATR in the laboratory experiment was maximized during the summer. Thus tail loss in the spring may not be due to male–male aggression as salamanders surface from overwintering fossorial retreats but perhaps to increased predation and/or female aggression during spring.

Wise and Jaeger (2016) concluded that seasonal variation in agonistic territorial behavior does occur within *P. cinereus* and that salamanders from the population at Hawksbill Gap are more aggressive than those at MLBS. We note that at MLBS, *P. cinereus* confronts no same-size congener while Hawksbill Gap is only a few meters below the eastern edge of the source talus where *P. shenandoah* resides and where it and *P. cinereus* compete for territories in the area of parapatry (see Griffis & Jaeger, 1998, in section 6.1, and Jaeger, 1970, in section 2.1). Jaeger had previously found some individuals of *P. shenandoah* at Hawksbill Gap amidst a dense population of *P. cinereus*.

If interspecific competition has led to increased intraspecific aggression by *P. cinereus* at Hawksbill Gap, then we pose two alternative hypotheses:

> H_1: Its increased aggression is a consequence of plasticity in aggressive behavior based on prior learning through combat with the highly aggressive *P. shenandoah* (Griffis & Jaeger, 1998, in section 6.1). If this were true, then (we posit) enhanced intraspecific aggressive behavior should decrease rapidly with increasing distance of *P. cinereus* from the source talus.
>
> H_2: Increased aggressive behavior by *P. cinereus* is genetically determined (e.g., through alpha selection; see section 6.6). If this were true, then intraspecific aggression should decrease more slowly with distance from the source talus, because the geographic spread of genes should be broader than learned one-on-one behavioral interspecific interactions would allow.

These hypotheses are ripe for future evolutionary ecologists to explore.

3.10.2 Geographical variation

For geographical variation, Wise and Jaeger (2016) noted that populations of *P. cinereus* occur over a vast range of latitudes and elevations (e.g., coastal lowlands to tops of mountains). In northern lowlands, *P. cinereus* occurs but often cannot forage on the forest floor during summers due to high temperatures there; however, populations on cooler mountains can forage during summers. Also, Hass (1985) reported geographic protein (genetic) variation among four groups of *P. cinereus* in the southern part of its range: Group I in the southern Blue Ridge Physiographic Province, Group II in the southern Appalachian Plateau, Group III in the northern Blue Ridge Mountains, and Group IV in West Virginia northeast of the New River (see Wise, 1995, for a fuller description of these groups). Therefore Wise and Jaeger hypothesized that latitudinal, altitudinal, and genetic variations will each result in territorial variation in behavior among populations.

Wise and Jaeger (2016) collected adult males of *P. cinereus* from eight locations: (1) western and (2) eastern Maryland; (3) Hawksbill Gap and (4) nearby Skyline Drive, both in northwestern Virginia; (5) MLBS and (6) Smyth-Grayson both in southwestern Virginia; (7) one site in West Virginia; and (8) one in Indiana. These locations were chosen to obtain populations from genetic Groups II (locations 2, 5, 6, 7) and III (1, 3, 4, 8); from low (1, 2, 7, 8) and high (3, 4, 5, 6) elevations; and with a congeneric species present (3, 6, 7, 8) or absent (1, 2, 4, 5). The congeners were *P. dorsalis, P. richmondi, P. shenandoah*, and *P. welleri*. All eight populations were collected during late September to mid-October 1993 and tested at ULL from early November to early December 1993 in a completely randomized design. Resident/intruder behavioral tests used the same experimental procedures and statistical tests as described earlier in section 3.10.1.

EDGE and ATR were negatively correlated with elevation, but this differed for residents and intruders. Residents from higher elevations (850–1,490 m) spent significantly less time in EDGE than did residents from lower elevations, but intruders exhibited no significant differences among elevations. However, intruders spent significantly more time in EDGE than did residents from all elevations. Thus, regardless of elevation, intruders were more submissive behaviorally than were residents, as predicted of resident advantage in territorial contests (Maynard Smith & Parker, 1976). Elevations did not significantly affect ATR by either residents or intruders, although intruders showed a trend toward more time in ATR at elevations 850 m and above. Overall, elevation had a significant influence on territorial conflicts by *P. cinereus*.

Populations in Hass's genetic Group II spent significantly more time in EDGE than did conspecifics in Group III. Neither residents nor intruders differed significantly in time devoted to ATR. Therefore, genetic differences between Groups II and III may influence territorial behavior (e.g., level of submission) by *P. cinereus*, but the authors cautioned that confounding differences among locations (e.g., social environments) may be associated with Hass's Groups II and III.

Wise and Jaeger (2016) were surprised to find no significant differences for times spent in ATR or EDGE for residents or intruders based on sympatry or allopatry with same size congeners. By contrast, the "seasonal variation" experiment had detected that *P. cinereus* at Hawksbill Gap (where *P. shenandoah* is a competitor) was significantly more aggressive toward conspecifics than it was at MLBS (where no same-size congener occurs). Yet, the "geographical variation" experiment found no such differences in aggression or submission for *P. cinereus* where *P. dorsalis, P. richmondi, P. shenandoah*, and *P. welleri* were sympatric with *P. cinereus* versus the four localities where no same-size congeners occurred. The authors posed a posteriori alternative hypotheses to explain these conflicting results: (1) the geographical variation experiment had been conducted in late autumn when the seasonal variation experiment had found the lowest levels of aggression by *P. cinereus,* and (2) different species of *Plethodon* vary in aggression toward *P. cinereus*. For example, *P. shenandoah* and *P. cinereus* are very aggressive (ATR and bites) toward each other (as related in section 6.1) while *P. hoffmani* demonstrates very little aggression toward *P. cinereus* (as in section 6.6). *Plethodon hubrichti* and *P. cinereus* engage in intermediate levels of aggression (as in section 6.7). Therefore, if interspecific aggression/submission carry over into intraspecific behavior by *P. cinereus*, vast differences may occur among sympatric localities, leading to the nonsignificant results in the geographical experiment. Future research should explore more carefully the agonistic interactions between *P. cinereus* and *P. dorsalis, P. richmondi, P. welleri*, and a host of other congeners.

3.11 SELECTED RECENT RESEARCH BY OTHERS: INTRASPECIFIC TERRITORIALITY

No comprehensive studies of geographical variation in the behavior of *P. cinereus* have been conducted since Wise's (1995) work. However, because this species provides a geographically widespread model system with which to test hypotheses regarding agonistic and territorial behavior, numerous researchers have examined behavior in various portions of the range. For example, Rollinson and Hackett (2015) described aggressive and spacing behavior of individuals of *P. cinereus* from near the northern limit of the species range (North Bay, Ontario, Canada) and reported that, in laboratory contests, individuals from these populations exhibited moderate degrees of aggressive behavior and a clear residency effect. Field surveys suggested, however, a random rather than uniform distribution of individuals. As discussed previously, Quinn and Graves (1999; see section 3.7) determined that, compared to individuals from MLBS, individuals from northern Michigan are more likely to aggregate in groups and are less likely to exhibit aggressive behaviors in laboratory trials. It may be that of *P. cinereus* in northern populations do not enter into competition for prey because, even though their activity is shortened by a compressed growing season, the forest floor remains continually damp during this time. The benefits derived from territorial defense, especially the exclusive access to prey under cover during dry periods, simply may not pay off in these populations.

In contrast, Anthony and Pfingsten (2013) summarized five studies on populations in northeastern Ohio (see their Table 18-3) and found that *P. cinereus* from that region exhibited ATR at levels on par with populations from

New York and Virginia. Individuals in one Ohio population in the Cuyahoga Valley National Park appear to exhibit spacing patterns that are similar to those described at MLBS. Of 518 adult salamanders observed beneath cover objects, 80 were paired with another adult, but only 3 same-sex pairs (all females) were observed (Anthony et al., 2008). The remaining 438 salamanders were found alone. Additionally, resident salamanders from this population, when removed from cover, were replaced by smaller "floater" individuals (Anthony & Pfingsten, 2013), a result that mirrors Mathis's (1991b) findings at MLBS. Similarly, Moore et al. (2001) reported that individuals of *P. cinereus* from the Allegheny National Forest of northwestern Pennsylvania are significantly smaller when found in litter compared to those found under rocks. In this population, salamander size was positively correlated with cover object size such that the largest and heaviest salamanders were found under the largest rocks, a result that suggests that only the largest individuals are able to exclude territorial competitors from contested cover. Positive correlations between SVL and cover object size have also been detected at MLBS (Mathis, 1990a) and in one northeastern Ohio population (Hickerson et al., 2004).

Ontario, Michigan, northern Ohio, and northwestern Pennsylvania all fall within geographic province IV described earlier (see section 3.10). At the last glacial maximum, these areas would have been inhospitable to salamanders and so must have been colonized relatively recently by *P. cinereus*. Despite this, considerable variation in intraspecific territorial behavior exists. These differences might reflect undiscovered genetic structure resulting from differences in the timing and routes of dispersal by populations of *P. cinereus*. Alternatively, geographical variation in behavior can emerge from a variety of other factors unrelated to genetics. For example, Maksimowich and Mathis (2000) explored the effects of parasites on territorial behavior in a related species of *Plethodon* (*P. angusticlavius*) and found strong effects of parasite load on behavior. Parasitized salamanders were less efficient at foraging for prey and less aggressive overall in territorial contests. Parasite load represents one variable that is rarely measured, but it has the potential to contribute to observed variation in behavior. Prey availability is yet another such variable. For example, Maerz and Madison (2000) compared territorial attributes of two populations in south-central New York that differed in their access to prey. At their "high-food" site, individuals of *P. cinereus* were less territorial. These individuals cohabitated with more same-sex individuals and were less likely to be associated with a fixed site compared to individuals from a "low-food" population.

Because the quantity and quality of prey are important factors during territorial acquisition and defense, we next examine the "rules" used by *P. cinereus* while foraging during periods of prey scarcity and prey abundance.

4

Foraging tactics by *P. cinereus* within territories

During 1968, Jaeger (1972) examined the stomach contents ($N = 1{,}820$ prey) of both *P. cinereus* ($N = 190$ salamanders) and *P. shenandoah* ($N = 36$) across days when the leaf litter in the area of sympatry varied in moisture. He found that the mean number of prey per salamander's stomach was inversely proportional to the time since last rainfall: at 7 days post-rain, the mean number of prey was 1.9 per stomach; at 3 days post-rain the mean was 4.6; and at <1 days post-rain the mean was 20.1 ($p < 0.001$). Stomachs were engorged with prey shortly after a rain, when salamanders could forage in the leaf litter (e.g., for collembolans), but stomachs were nearly empty when the leaf litter dried. In the latter case, salamanders were largely confined to patches of moisture under rocks and logs or in underground crevices. Jaeger concluded that prey are periodically a limited resource for which competition would occur, and this was supported by later research reviewed in chapter 3.

This hypothesis of food limitation led to a series of laboratory experiments to learn how red-backed salamanders forage within their territories, using foraging theories (e.g., Krebs, 1978) as a theoretical foundation for the research. The experiments focused on "optimal choice of diet" (e.g., Estabrook & Dunham, 1976), because this theory dealt with foraging tactics that "should" vary from times of food abundance to periods of food scarcity.

4.1 FORAGING ON LIVE VERSUS DEAD PREY

David and Jaeger (1981) began by observing *P. cinereus* foraging on live versus recently dead (frozen then thawed) prey (*D. melanogaster*), because dead flies were easier for the authors to handle. Also, because red-backed salamanders mostly forage in the dimly lit leaf litter or at night while climbing plants, chemical detection of prey may be more important than visual detection of prey.

David and Jaeger (1981) conducted three laboratory experiments and a set of observations on Hawksbill Mountain to test the significance of four hypotheses.

H_1: Red-backed salamanders will show a stronger feeding response toward dead flies compared to moving live flies.

H_2: Previous short-term experience with live flies will enhance the subsequent feeding response toward dead flies.

H_3: Previous experience with a particular prey type will enhance the subsequent feeding response toward that prey-type (i.e., development of a "search image").

H_4: Dead flies will be detected primarily via chemical cues.

Experiment 1 tested H_1 and H_2. Equally hungry salamanders ($N = 21$ not fed for 7 days) were individually tested randomly in each of four treatments: (1) five live flies blown into the chamber during both the first and second 15-minute feeding bouts; (2) similarly, five live flies then five dead flies; (3) five dead then five live flies; and (4) five dead flies in both 15-minute bouts. The salamanders significantly preferred live to dead flies during the full 30-minute tests, which did not support H_1. Consequently, dead flies were seldom used in future experiments. Hypothesis 2 was not supported, because previous exposure to live flies did not significantly enhance feeding responses to dead prey.

Experiment 2 tested H_3. Naïve salamanders ($N = 21$) were given five live flies during both the first and second 15-minute feeding bouts, while 21 different naïve salamanders were given five dead flies both times. These data were compared with the feeding responses of salamanders ($N = 21$) when encountering either just live or just dead flies during the first and second feeding bouts. The data supported H_3, because experienced salamanders ate significantly more flies per 30 minutes (both live and dead ones) than did the naïve (no previous experience) salamanders. This suggests that *P. cinereus* can learn to forage more successfully (in rate of prey captures) through prior learning experiences with particular types of prey: that is, they can develop a search image through learning. We explore learning by *P. cinereus* in more detail in chapter 9.

Experiment 3 tested H_4 concerning how dead prey are detected by *P. cinereus*. Each of 21 salamanders was randomly tested for feeding response in each of four treatments: (1) no light or odor sensory deprivation when dead flies were presented; (2) visual deprivation (flies encountered in a completely dark chamber); (3) chemical cue deprivation (in a lit chamber but the flies were coated with a transparent gel to inhibit chemical cues); and (4) gel-coated flies in a completely dark room. The data significantly supported H_4; that is, the salamanders required chemical cues but not visual cues to locate dead (immobile) prey.

David and Jaeger's (1981) data from Hawksbill Mountain comprised ~9,500 prey ingested by *P. cinereus*. Only 1.8% of their diet by number and 5.0% of diet by volume of prey were composed of nonmoving prey (insect pupae). Stationary prey thus comprised only a small, but perhaps important, fraction of the diet. This implies that visual detection of mobile prey is of major importance to *P. cinereus* in the forest.

4.2 DIET BREADTH

Salamanders are (or were) often said to be "indiscriminate feeders" in that they would eat any prey item that could be captured on the tip of their tongue (smaller prey) or grasped by their jaws (larger prey). For example, Jaeger (1990) showed that *P. cinereus* on Hawksbill Mountain ingested the following taxa: Nematoda, Gastropoda, Oligochaeta, Acarina, Araneae, Isopoda, Chilopoda, Diplopoda, and 10 orders of Insecta. However, the Acarina (mites) comprised 50.02% of the diet by number (i.e., slowly moving prey with highly chitinous exoskeletons) while Collembola (springtails) and Diptera (flies) (i.e., faster-moving prey with softer exoskeletons) comprised 13.09% and 20.85% of the diet, respectively. Jaeger argued that *P. cinereus* "preferred" (fed discriminately on) prey with softer exoskeletons (e.g., collembolans) because such taxa would be easier and faster to digest, thus providing more assimilated energy and nutrients per unit time of foraging. Gabor and Jaeger (1995, in section 3.8) later tested this assumption. Jaeger and Lucas (1990) also postulated that such foraging decisions (for any taxon) have an evolutionary foundation, in that the evolution of foraging decisions (e.g., choice of diet) should be based on the average fitness payoffs of alternative foraging behavioral patterns.

4.3 OPTIMAL PREY CHOICE

Beginning with MacArthur and Pianka (1966), many theories were proposed concerning the evolution of foraging tactics, but at the time empirical tests of the theories were rare and largely confined to birds (see Pyke et al., 1977, for a review of then-known theories). Jaeger and Barnard (1981) were especially interested in the theory of optimal prey choice, which depends on the predator's ability to distinguish among prey of varying profitabilities and to choose the more energetically profitable types during foraging. Theoretically, the predator's net energy gain should depend on the profitability of the prey types included in the diet and the energy expended while searching for those prey types (which varies with prey density). When the most profitable prey type is abundant and easily found, so energy expended during the search is low, the predator should maximize net energy gain by specializing on that type alone. Yet as the abundance of that prey type decreases, the predator should expand its diet to include the next most profitable prey type and do so in sequence until all prey types are scarce, at which point the predator should ingest all available prey types that yield a positive value for profitability. In short, the predator should be a specialist feeder on the best prey type when it is readily available, but when that prey type decreases in density, it should become more of a generalist feeder.

This theory had obvious implications for salamanders on Hawksbill Mountain that were confronted with abundant prey (of many taxa) during and shortly after a rainfall but found fewer and fewer prey as the leaf litter dried post-rain (Jaeger, 1972). Therefore, Jaeger and Barnard (1981) tested whether *P. cinereus* would shift from a specialized to a generalized diet as prey densities decreased in the salamanders' territories.

Red-backed salamanders (*N* = 24 collected in Albany County, New York) were allowed to establish territories in separate chambers and were presented with two prey types at varying densities: the larger *D. virilis* (L) and the smaller *D. melanogaster* (S). The salamanders were individually and randomly presented with the following densities of flies: 4L:4S, 10L:10S, 16L:16S, 22L:22S, and 16L:32S. The first four densities tested for optimal choice of diet, and the last density, when compared to 16L:16S, tested for choice of diet based on absolute or relative densities of prey. To keep densities constant within each test, when an L or S fly was eaten by a salamander, another L or S fly was blown into the test chamber via a plastic tube.

Jaeger and Barnard (1981) recorded the following foraging behavioral data: (1) intercapture intervals; (2) number of tongue strikes before either capturing or giving up on a prey item; (3) the sequence of L and S flies ingested; (4) whether a salamander pursued an L or S fly or lay in ambush for it; and (5) whether the salamander ignored an encountered L or S fly, struck at it and missed, or ingested it. They statistically compared the salamanders' responses among the various test densities, but we report only significant differences. They found that *P. cinereus* had an indiscriminate diet at low prey densities but specialized on L flies at higher densities. Specialization occurred by increasingly ignoring encountered (i.e., within tongue-striking distance) S flies but not ignoring L flies. These data generally supported the central tenet of the theory of optimal diet choice and showed that foraging decisions by *P. cinereus* vary with prey densities.

The salamanders exhibited partial prey preferences at high densities, because S flies were not completely eliminated from their diet. Choice of diet was not based strictly on the absolute abundance of L flies (contrary to theory), perhaps because the salamanders made mistakes in judging the profitability of S flies at high densities when flies were captured rapidly. Such mistakes are suggested by a higher proportion of aborted attacks on S flies than on L flies at higher densities. The salamanders switched from pursuit of flies to ambush tactics with increasing density, and capture rates were higher by ambush than by pursuit. Apparently the salamanders did not use capture rates to monitor prey densities but instead used prey encounter rates.

Using published estimates of the caloric contents of L and S flies, the digestive assimilation efficiencies from Bobka et al. (1981) and the estimated energy budget of *P. cinereus* (Merchant, 1970), Jaeger and Barnard (1981) developed a mathematical model to assess each salamander's net energy gain (E) at various densities of flies. (The mathematical model is presented in Jaeger & Rubin, 1982, section 4.6.) Unfortunately, the original model predicted that, the more flies eaten by a salamander, the less its net energy gain—until the salamander would eventually disappear from overeating! Luckily, this model was corrected before submitting the manuscript for publication, but this suggests that Jaeger and Barnard as well as salamanders can make dumb mistakes. The subsequent published model suggested that net energy gained from foraging by the salamanders increased from low to high prey densities as a result of a concurrent increase in specialization on L prey (increased assimilation efficiency) and increase in ambush foraging (decreased energy expended while foraging). In sum, *P. cinereus* seems to be very efficient in its foraging decisions concerning both choice of diet and its own alternative behavioral tactics while foraging.

4.4 TERRITORIAL AND FORAGING BEHAVIORAL CONFLICTS

Jaeger et al. (1981) next asked if territorial behavior and optimal diet choice (as in Jaeger & Barnard, 1981) might be in conflict with each other for *P. cinereus*. Salamanders ($N = 22$) were allowed to establish territories in test chambers and were randomly manipulated among four conditions: (1) each salamander was left undisturbed in its home chamber (the control condition); (2) each was disturbed by moving it to another clean chamber and then back to its pheromone-marked home chamber; (3) each was disturbed as in (2) but its own pheromone-marked substrate was replaced by a clean (no pheromones) substrate; and (4) each was disturbed and its substrate replaced by one pheromone marked by another salamander. Then, for each condition, 22L:22S flies were blown into the test chamber. Data recorded were (1) number of L and S flies ingested, (2) number of encountered flies that were either eaten or ignored,

(3) intercapture intervals, (4) rate of the salamander's marking behavior (cloacal area or chin touched to the substrate), and (5) amount of time spent in the FLAT posture. Statistical tests among the four conditions yielded the following significant results.

The salamanders foraged optimally, specializing on L flies, when undisturbed on their own territorial pheromones (condition 1). However, they did not forage optimally on a clean substrate (condition 3) or on a conspecific-marked substrate (condition 4). In the latter two situations, their rate of net energy gain was reduced due to the failure to specialize on L prey and their longer intercapture intervals while time was devoted to either marking behavior or submissive posturing (FLAT). Also, just disturbing the salamanders (condition 2) slightly, but not significantly, altered their foraging behavior. Jaeger et al. (1981) suggested that the salamanders opt for a lower caloric yield from foraging until they have established their own pheromone-marked territories, after which they switch to a higher sustained caloric yield by optimally foraging. In brief, foraging tactics and territorial behavior (at least for disturbances and substrate pheromones) are indeed in conflict for *P. cinereus*.

4.5 ASSESSING PREY DENSITIES

Jaeger, Barnard, et al. (1982) next asked how *P. cinereus* assesses prey densities while optimally foraging: for example, by encounter rates or capture rates of prey, or by some numerical means such as counting available prey? The ability of a predator to assess prey densities is crucial to optimal foraging theories and to the foraging behavior of red-backed salamanders (Jaeger & Barnard, 1981).

Jaeger, Barnard, et al. (1982) designed a series of experimental conditions that were intended to vary the ability of the salamanders to perceive visually (i.e., encounter) prey while holding constant the densities of L and S flies and capture rates. The experimental design was the same as in Jaeger and Barnard (1981) except that barriers were added under three of the four testing conditions (C) in the salamanders' home (territorial) chambers: C_1: no barriers (the control), C_2: four transparent glass microscope slides in each chamber, C_3: eight transparent slides, and C_4: four opaque slides in place of the transparent ones. The slides were arranged such that both salamanders and L and S flies could move over and around them, but C_2 and C_3 increased habitat heterogeneity without inhibiting a salamander's ability to detect visually all flies in the chamber. Condition 4 restricted the salamander's visual perception to just the subdivision (between two black slides) of the chamber the salamander happened to occupy. Salamanders ($N = 22$) were tested randomly in each of the four conditions, and 22L:22S flies were blown into each chamber. The salamander was allowed to eat 10 flies before each test ended, and each eaten fly was replaced by another L or S fly (thus densities of L and S flies remained constant). The salamanders' behavioral responses were monitored for (1) number of L and S flies eaten, (2) intercapture intervals, and (3) the number of L and S flies encountered (i.e., came within striking distance of the salamander's tongue) and were either eaten or ignored (not eaten). Statistical tests were conducted among the four test conditions.

In C_1 (no slides), the salamanders foraged optimally (replicating Jaeger & Barnard, 1981) by specializing on L flies at such a high prey density. Significantly more L than S flies were eaten in C_1 through C_3 (vision not inhibited) but not in C_4 (visual inhibition). Also, encountered L flies were ingested significantly more than encountered S flies in C_1 through C_3 but not in C_4, because more encountered S flies were ignored when transparent barriers were present than when opaque barriers hindered visual perception of flies. Intercapture intervals (i.e., capture rates) remained constant among compared conditions, suggesting that the number and type of barriers did not hinder the movement of either flies or salamanders in the chambers.

Jaeger, Barnard, et al. (1982) concluded that the salamanders used visual encounter rates (not capture rates) to assess prey density, and they speculated that salamanders "counted" the number of prey items within the visual field. However, Uller et al. (2003) later disputed the notion that salamanders "count" prey, and they suggested instead that *P. cinereus* uses "numerical discrimination" of prey numbers to determine prey densities, as described in section 9.1.

Why would *P. cinereus* employ encounter rates to assess prey densities when capture rates would provide additional information concerning the costs of pursuing and handling (ingesting) prey (which are important parameters in theories of optimal choice of diet)? Jaeger, Barnard, et al. (1982) suggested that these costs are energetically very low for *P. cinereus*; prey are pursued for only a short distance, and handling times are energetically inexpensive because a single tongue-flick captures one to three prey items and they are swallowed rapidly and intact. Also, encounter rates with prey would be faster in assessing changes in prey densities than would slower capture rates. Of course it is unknown whether the use of encounter rates is inherited (encoded in the salamanders' DNA) or learned from prior experiences of foraging (all of the behavioral tests described here were with adults) or an interaction of the two. What is evident from this research, though, is that *P. cinereus* is quite skilled in detecting prey densities.

Hill et al. (1982) tested the hypothesis that *P. cinereus* can learn and remember the locations of aggregations ("patches") of prey: dead L and S *Drosophila*. This was a preliminary experiment ($N = 8$ salamanders), so we present only their conclusions. Over 28 days of foraging among patches of flies, the salamanders failed to learn to distinguish patches containing L or S flies. The authors suggested that, in natural forested habitats, invertebrate prey may not be distributed in discrete, long-lasting aggregations. If this were true, then natural selection would not be expected to select for "patch foraging" behavior (*sensu* Krebs, 1978; Smith & Dawkins, 1971). As described in section 9.3, Gibbons et al. (2005) later found that learning does play a role in some aspects of foraging behavior in *P. cinereus*, but patch foraging may not be one of them.

4.6 JUDGING PREY PROFITABILITIES

A predator's ability both to assess and to rank (by their profitabilities) all potential prey types is another critical tenet for theories of optimal choice of diet, and such ability is evident for *P. cinereus* as shown by Jaeger and Barnard (1981). One simple decision rule might be that bigger prey types are better than smaller ones, and in Jaeger and Barnard's experiment salamanders specialized on L flies compared to S flies at high prey densities (as theory predicted). However, for *P. cinereus* with a diet of numerous invertebrate taxa, "bigger is better" might be a poor rule to use. For example, the larger snails and hymenopterans in its diet are more heavily armored than the smaller, lightly armored collembolans, and the stomach contents of *P. cinereus* on Hawksbill Mountain contained only 2.02% gastropods and 5.44% hymenopterans compared to 13.09% collembolans (Jaeger, 1990). Therefore, an important factor for *P. cinereus* would not only be the size of the prey ingested but its rate of digestion (see Gabor & Jaeger, 1995, in section 3.8). More heavily armored prey may "clog" the digestive tract compared to lightly armored prey due to (1) longer digestive times in the stomach, (2) poorer assimilation efficiencies (Bobka et al., 1981), and (3) longer passage times through the digestive track (mouth to cloaca). So how does *P. cinereus* distinguish among the caloric (and nutritional) profitabilities among >100 potential invertebrate prey types? Jaeger and Rubin (1982) attempted to distinguish among three alternative hypotheses:

H_1: By innate recognition via evolution with those prey types.

H_2: By the sizes of the prey types (i.e., "bigger is better").

H_3: By learning through previous experiences from eating those prey types.

Salamanders from the Catskill Mountains of New York were trained for three months in their home (territorial) chambers, with 15 *P. cinereus* each randomly assigned to only one of four diet conditions: C_1: both L and S flies (as in previous experiments), C_2: L flies only, C_3: S flies only, and C_4: a nondipteran prey (coleopteran larvae and adults of *Tribolium castaneum*). After the training period, the prey were removed and each of the 60 salamanders had 22L:22S flies blown into their chambers.

During the foraging tests, Jaeger and Rubin (1982) blew in an L or S fly for each one eaten to keep prey densities constant. They recorded (1) the number of L and S flies eaten, (2) the number of each that was encountered and either ingested or ignored, (3) the number of each captured by pursuit or by ambush, (4) the number of strikes with the tongue required to capture an L or S fly, (5) intercapture intervals for L and S flies, and (6) the number of seconds that each salamander moved and was motionless. They also estimated the rate of E for each salamander during a foraging test, which ended after 10 flies were eaten or, if the salamander failed this criterion, after 45 minutes of waiting (for only 2 of the 60 foragers). The equation was

$$E = \frac{a(Lb + Sc)}{t} - \frac{W(xk_1 + yk_2)}{t}$$

where a = assimilation efficiency of *P. cinereus* for *Drosophila* at 15°C (mean = 0.86 from Bobka et al., 1981); L = number of L flies ingested; b = caloric value of an L fly (mean = 3.0 cal from Jaeger & Barnard, 1981); S = number of S flies ingested; c = caloric value of an S fly (mean = 1.2 cal); t = time to eat 10 flies; W = dry mass of each salamander measured postmortem; x = total time moving by each salamander; k_1 = metabolic cost of movement (mean = 5.764×10^{-3} cal/minute/g body mass); y = total time at rest by each salamander; and k_2 = metabolic cost at rest (mean = 4.549×10^{-3} cal/minute/g, with both k_1 and k_2 from Merchant, 1970). These values of E (for each salamander in each condition) are important in estimating the energetic profitabilities of foraging among conditions 1–4. For all variables, statistical tests were performed between conditions.

The significant results showed that only in C_1 (trained on both L and S flies) did the salamanders forage optimally (specializing on L flies). They specialized by (1) rejecting more encountered S than L flies, (2) pursuing (rather than ambushing) more L than S flies, (3) having shorter intercapture intervals with L flies, and (4) having higher positive rates of E. These results did not support the hypothesis that *P. cinereus* has inherited abilities to assess and rank the profitabilities of different prey types, because in C_1 the salamanders specialized on the more profitable (larger) *D. virilis* even though this species does not occur in its natural, forest habitats (Jaeger, 1972, 1990). Also, the results did not support the hypothesis that the salamanders used the "bigger is better" decision rule, because they did not specialize after training in C_3 and C_4 (i.e., with S flies only vs. much larger coleopterans only).

Therefore, Jaeger and Rubin (1982) concluded that the salamanders learned through foraging experiences (with L and S flies in C_1) to assess and rank profitabilities of the prey types. They suggested that learning may be necessary to assess both (1) the gross caloric (and possibly nutritional) value of different prey types and (2) the rate at which those prey types can be digestively assimilated once ingested, which would vary with the chitinous content of invertebrate prey. Note that in all testing conditions, the L flies had proportionally less chitin in the exoskeleton than did S flies, based on their surface to volume ratios.

Of course Jaeger and Rubin's (1982) experiment did not completely eliminate the alternative hypotheses that *P. cinereus*, in some cases, uses "innate knowledge" of profitabilities or uses the "bigger is better" rule. The weakness of this experiment was that no natural prey (inhabiting the forest with *P. cinereus*) were used in any of the four training or testing conditions. A better (but far more difficult) test of the learning hypothesis would be to test foraging decisions, over months, by *P. cinereus* from naïve, neonate age to later older ages with more learning experiences. Using natural prey would be preferable. The learning hypothesis would be supported if neonates foraged as generalists at high prey densities of alternative prey while, concurrently, older ages increasingly specialized on the more profitable prey type with experience in foraging. No one has conducted such a laborious experiment.

4.7 CONFLICTS BETWEEN FORAGING BEHAVIOR AND TERRITORIAL DEFENSE

Jaeger et al. (1981), in section 4.4, showed how territorial disturbances and the presence of pheromone-marked substrates from conspecifics can detract from optimal foraging behavior of territorial *P. cinereus*. Later, Jaeger et al. (1983) examined the major topic of how territorial intrusions by live conspecifics might impede such foraging tactics by territorial residents. In their experiment, the same experimental design was used and the same data from the residents recorded as in Jaeger and Rubin (1982) in section 4.6. Net energy gain (E) was estimated separately for each salamander in each of the six conditions using the equation in Jaeger and Rubin (1982).

Salamanders established territories in individual home chambers and fed on L and S flies during the pretest period. Then each salamander ($N = 22$) was tested randomly in each of the six conditions: C_1: resident not subjected to any of the following impingements into its territory (general control); C_2: a damp paper towel in the shape of a *P. cinereus* placed in each chamber (surrogate control); C_3: a surrogate infused with pheromones of a previously familiar conspecific placed in the chamber (familiar surrogate test); C_4: a surrogate infused with pheromones of an unfamiliar conspecific (unfamiliar surrogate test); C_5: a previously familiar live intruder introduced (familiar intruder test); and C_6: an unfamiliar live intruder introduced (unfamiliar intruder test). (See section 9.5 for how *P. cinereus* learns to distinguish familiar from unfamiliar conspecifics.)

Then, for each condition, 22L:22S flies were blown into each chamber, each eaten fly was replaced, and the test ended when a salamander had ingested 10 flies. The general hypotheses by Jaeger et al. (1983) were that intruder threat would increase sequentially from C_1 to C_6. Consequently, the territorial residents would, in rank order, subtract time from foraging due to increasing territorial defense and, because of this, *E* would decrease from C_1 to C_6.

The statistical results are too complex to review in detail here, but a general summary follows (from Jaeger et al., 1983). As the degree of competitive threat increased from C_1 to C_6, more time was devoted to territorial defense via postural displays (all trunk raised and FLAT) and biting of live intruders, at the expense of foraging. Also, territorial residents gradually shifted from a specialized diet on L flies to an indiscriminate diet, even though prey densities and encounter rates with L and S flies did not change. The unfamiliar pheromones (C_4) and both types of live intruders (C_5 and C_6) led to ~50% decrease in residents' rates of *E*, of which ~80% was due to time withdrawn from foraging and ~20% was due to change in diet (from specialist to generalist foraging).

The results of this experiment suggests that the decrease in *E* during territorial defense is a major cost of territoriality for *P. cinereus*. This would especially apply in forests where salamander densities are very high and territorial intruders are common (as at our research sites in Virginia: Gillette, 2003; Mathis, 1990a, 1991b).

Research concerning optimal (and sometimes suboptimal) choice of diet ended here, because after two years of counting L and S flies and watching, for hundreds of hours, the behavioral tactics of *P. cinereus*, Jaeger and his five undergraduate students (all coauthors) were exhausted. Still, many more hypotheses are yet to be tested. One hypothesis is that adult males and females of *P. cinereus* forage and defend differently. Females may need to maximize *E* from foraging in order to yolk eggs, usually about six to seven laid per female every second spring at Mountain Lake Biological Station (e.g., Gillette, 2003; Peterson, 2000). Males may devote more time to territorial defense with sacrifices to *E*.

Thus far we have reviewed assumptions that pheromones are involved in inter- and intraspecific communication (section 2.2 and chapter 3, respectively) and that they influence foraging decisions (this chapter). These assumptions concerning unseen pheromones are inductive (Hume, 1748), because they merely reflect seen changes in behaviors by *P. cinereus* based on presumed, unseen odoriferous markings on substrates. In chapter 5, we turn to more definitive studies that directly concern the various signals that pheromones communicate and, more important, their glandular sources.

4.8 DIET DIVERSITY AND CLUTCH SIZE

Jaeger (1981c) was curious as to why fully terrestrial salamanders (e.g., *Plethodon*) appear to have smaller and less frequent clutches of eggs than do aquatic salamanders (e.g., *Desmognathus*). He speculated that terrestrial salamanders should forage under variable conditions of moisture in the leaf litter (rainy to dry periods), which may lead to a large diversity of prey taxa in their diets (i.e., generalist foraging tactics). By contrast, aquatic salamanders should forage under less variable conditions in their streams and ponds, which may lead to a smaller diversity of prey taxa in their diets (i.e., specialist foraging tactics). He wondered if these differences in moisture and presumed foraging tactics may have led to natural selection for differential reproductive tactics. For example, species of *Plethodon* usually lay a small clutch of large eggs biennially while species of *Desmognathus* usually lay a larger clutch of smaller eggs annually, or so it appeared to Jaeger from his observations at Mountain Lake Biological Station. So he wondered if a possible linkage occurs between the life-history traits of diet diversity (foraging tactics) and clutch size/frequency (reproductive tactics).

The easiest approach for a preliminary survey was a search of the literature. Jaeger (1981c) compared the diet breadths of terrestrial and aquatic species. He grouped the ingested prey into taxonomic orders for Insecta and Arthropoda and into classes for all other prey. From these data, he computed diet diversity (H') indices (Shannon & Wiener formula) for 19 terrestrial and 21 aquatic species in North America. The terrestrial species were in the families Plethodontidae (N = 15 species), Ambystomatidae (adults, which breed in ponds but forage in the forest leaf litter, N = 3), and Salamandridae (the terrestrial eft stage of *Notophthalmus viridescens*, N = 1). The aquatic species were in Salamandridae (N = 11), Ambystomatidae (N = 4, larval stage), Plethodontidae (N = 2), Dicamptodontidae (N = 1, larval stage), Amphiumidae (N = 1), Cryptobranchidae (N = 1), and Sirenidae (N = 1). He also compared the clutch sizes and frequencies of terrestrial (N = 7 species) and aquatic (N = 7) Plethodontidae. He chose this single family to avoid differential selection for life-history traits among multiple families. Within Plethodontidae, he compared size-matched terrestrial versus aquatic species to reduce differential clutch sizes among females of different body sizes (i.e., larger females tend to lay more eggs per clutch than do smaller conspecifics; e.g., Wise, 1995).

Jaeger (1981c) found that terrestrial species had larger prey diversity values (mean H' = 2.44, range 1.49–3.12) than did aquatic species (mean H' = 1.62, range 0.29–2.46), and the differences were significant ($p = 0.00005$). Focusing on just two species that are compared behaviorally in sections 6.3 and 6.4, H' for *P. cinereus* was 2.08 to 2.18 (varying among the three studies in the literature) while H' for *Desmognathus fuscus* could not be computed from the literature in 1981. Clutch sizes for terrestrial Plethodontidae ranged from a mean of 5 eggs/clutch (range 3–80) for *P. hoffmani* to a mean of 26 eggs/clutch (range 16–34) for *P. glutinosus*. For aquatic species, clutch sizes ranged from a mean of 16 eggs/clutch (range 8–25) for *D. ochrophaeus* to 126 eggs/clutch (range 77–192) for *Pseudotriton (*now *Plethodon) montanus*. For *D. fuscus*, the mean was ~ 31 eggs/clutch (range 18–45) compared to a mean of only 12 eggs/clutch (range 8–16) for *P. cinereus*. The breeding cycle for terrestrial species was biennial for four species of *Plethodon*, thought to be biennial for *P. shenandoah*, and unknown for *P. welleri* and *Aneides aeneus*. The cycle for aquatic species was annual for six species and irregular for *Pseudotriton montanus*. A dearth of information in the literature by 1981 prevented a more comprehensive survey.

Jaeger (1981c) inferred that terrestrial species of caudates tend toward generalist diets, probably due to varying moisture on the forest floor when foraging. For example, *P. cinereus* can forage in the leaf litter during wet days but is constrained to foraging under or near cover objects when the leaf litter dries. Therefore, they act as "pulse feeders" with engorged stomachs during wet periods and nearly empty stomachs during dry periods (Jaeger, 1972). By contrast, aquatic caudates tend toward specialist diets, probably because streams and ponds reduce the likelihood of severe fluctuations in moisture when foraging, even for streamside species. Therefore they should be able to forage nearly constantly, and this, according to optimal foraging theory (e.g., Emlen, 1966), should lead to selection for diet specialization.

However, an evolutionary connection between foraging tactics and reproductive tactics is dubious. Jaeger (1981c) noted that some theories (e.g., *r* and *K* selection *sensu* MacArthur & Wilson, 1967) proposed that such a connection should be an evolutionary solution. Animals living in food- or space-limited environments should have increased fitness by producing few young per reproductive cycle while animals living in environments with high juvenile mortality should have increased fitness by producing larger clutches per reproductive cycle. For example, there are many caudate predators found in streams and ponds such as fish, birds (e.g., egrets), crustaceans, and insect larvae (e.g., dragonflies). These predators are lacking in terrestrial forest environments. Despite this theory of "*r* and *K* selection," one should be dubious about a direct relationship between foraging tactics and reproductive tactics. Both may be under very different selection, such as differential predatory pressures. Therefore, the "relationship" found by Jaeger may be spurious.

4.9 SELECTED RECENT RESEARCH BY OTHERS: FORAGING TACTICS

An approach lacking in most field studies of diet and foraging are measures of prey selectivity by predators. This is because measuring available prey in the environment can be difficult. One approach is to use complete sampling techniques that extract all invertebrates from leaf litter samples surrounding a resident's territory. Paluh et al. (2015) used such an approach to compare the stomach contents of territorial individuals of *P. cinereus* to the prey available in the 1 m^2 area surrounding where salamanders were captured. They focused on one important group with which they had taxonomic expertise: ants. Three significant findings emerged from this study. First, salamanders were selective in their foraging, choosing ant species that were less aggressive and less likely to be chemically defended. Second, the striped morph of *P. cinereus* was found in more ant-rich territories and, perhaps as a result, consumed more ants than did the unstriped morph. This result suggests that striped salamanders may be more adept at securing more profitable territories (see section 10.1). Interestingly, many of the ants consumed by salamanders were not native species but were species introduced to North America.

Nonnative invertebrate prey make up an important component of the diets of *P. cinereus* in disturbed areas, suggesting that salamanders can learn to incorporate new taxa into their diets when other prey are unavailable or if novel prey species are more profitable. Maerz et al. (2005) examined diets of individual *P. cinereus* ($N = 2{,}009$) from six different forest plots in New York and Pennsylvania. They found that in lowland plots, nonnative earthworms make up the majority of prey by volume and argued that introduced prey, especially earthworms, augment rather than replace native prey in the diet of salamanders. This has interesting implications for territorial defense, because if salamanders are not limited by prey availability, the costs of territorial behavior may outweigh the benefits. Maerz et al. noted that salamanders from worm-invaded plots had denser and thicker tails and that females produced larger and more frequent clutches. As predicted, these salamanders also failed to exhibit site fidelity and spatial distributions consistent with holding territories.

The types of introduced prey species that individuals of *P. cinereus* incorporate into their diets vary considerably. For example, in Maerz et al.'s (2005) study, introduced worms made up an important component of the diet, but introduced ants were rarely eaten. Ivanov et al. (2011) found the opposite pattern in Ohio. A newly introduced ant dominated the diet, but introduced worms were rarely taken. These field studies suggest that individuals of *P. cinereus* can rapidly incorporate novel prey into their diets, but it is unclear what role learning, if any, plays in this process. The relationship of learning to behaviors that have stronger genetic underpinnings is of interest because heritable behaviors can evolve via natural selection while learned ones cannot. These relationships can be addressed in laboratory studies (see Gibbons et al., 2005, in section 9.2) where salamanders of known genetic relatedness can be tested for their ability to learn to recognize novel prey. Learning may also be important for identifying profitable patches of prey within which to forage. Crane and Mathis (2011) trained Ozark zigzag salamanders (*P. angusticlavius*) to associate prey with a landmark in experimental arenas. Trained salamanders spent twice as much time associated with landmarks than did members of a control group. In nature, retention of spatial information could be adaptive because individuals that can locate consistent prey patches, or refuges from predators, would be at an advantage compared to those that search randomly.

In chapter 5, we turn to the role that glands and pheromonal cues play in territorial advertisement and in social communication.

5

Pheromonal glands and pheromonal communication by *P. cinereus*

5.1 EARLY STUDIES SUGGEST PHEROMONES DO OCCUR

Jaeger (1986) reviewed the earlier studies that led to our later experiments concerning pheromones used during territorial defense and social communication. Four approaches dominated those earlier studies.

1. A test salamander is placed on a substrate half marked by another salamander and half marked by itself (or left as a blank control). If the test animals spend significantly more time on one half of the chamber than on the other, then those animals must have received some information concerning the previous presence or absence of another salamander's secretions (e.g., Jaeger & Gergits, 1979, as reviewed in section 2.2).
2. One salamander is allowed to mark a chamber, another is allowed to mark a second chamber, and a third chamber is left as a control blank. If certain behavioral responses (e.g., rate of feeding or nose tap [NT] with the nasolabial cirri) of the test salamanders vary significantly

among chambers, then the presence or absence of pheromones is inferred (e.g., Jaeger et al., 1981, 1983, reviewed in chapter 4).

3. A test salamander walks down a Y-tube in which the two air streams pass over two constrained salamanders, or over one salamander paired with a blank control, or over differently marked substrates. If the test animals exhibit a significant choice for one air source over the other, then pheromonal detection is inferred (e.g., Madison, 1975).
4. A test salamander is given a choice between substrates or burrows containing different sources of fecal pellets or cloacal washes. The sources might be (a) the test animal's own pellet versus one from a different salamander, (b) the test animal's pellet versus a surrogate (pellet-size wadded paper) control, or (c) a conspecific salamander's pellet versus a surrogate control pellet. If the test animals make significant choices between substrates (or burrows) based on type of pellet (or cloacal wash), then the presence or absence of pheromones is inferred (e.g., Jaeger et al., 1986; Simon & Madison, 1984).

Jaeger (1986) concluded that, apparently, pheromones are produced by salamanders of various species of Plethodontidae (as mentioned). However, little or nothing was known, in 1986, about (1) the glandular sources of these pheromones used in territorial or other social communications, (2) how males and females within a species might differ in such pheromones, and (3) what signals are transmitted to receivers by those pheromones. These topics stimulated a flurry of experiments, as described in this chapter.

Note that all of the experiments described in chapters 5 through 9 were conducted at Mountain Lake Biological Station (MLBS), Virginia, or the University of Louisiana at Lafayette with salamanders collected at MLBS, with two exceptions, which we mention later. We emphasize the salamanders' original locations because Wise and Jaeger (2016), in section 3.10, found that *P. cinereus* varies geographically in territorial behavior. The salamanders in our reviews varied in site of origin: several in northern New York, at Shenandoah National Park, and at MLBS.

5.2 DO MALES OF *P. CINEREUS* PRODUCE TERRITORIAL PHEROMONES?

Jaeger et al. (1986) tested the hypothesis that adult males produce pheromones, contained in fecal pellets, which identify male-marked territories. Each salamander ($N = 25$) was tested in a chamber under four randomized conditions: C_1: a burrow marked with its own fecal pellet versus another marked with a conspecific male's pellet; C_2: own-marked versus surrogate-marked (control) burrows; C_3: conspecific-marked versus surrogate-marked (control) burrows; and C_4: a general control of two surrogate-marked burrows.

Males spent significantly more time in their own-marked burrows than in conspecific-marked burrows, in surrogate-marked compared to conspecific-marked burrows, and in own-marked compared to surrogate-marked burrows. No position bias was found in the general control condition. Males also exhibited significantly more time NT (olfactory sampling: Fig. 3.1H) to conspecifics' pellets than to surrogate controls and spent significantly more time in the FLAT posture (Fig. 3.1 B) in front of the conspecific-marked burrows than in front of either their own- or surrogate-marked burrows. Jaeger et al. (1986) inferred that males of *P. cinereus* (1) mark their own territories with pheromones, (2) prefer their own marked shelters, and (3) either avoid or act submissively toward shelters of conspecific males. They also inferred that males use fecal pellets to mark (with odors) their territories, but this conjecture was later questioned by Simons et al. (1993, 1994). They showed that the postcloacal gland produces pheromones that can then be rubbed over deposited fecal pellets (see section 5.5).

5.3 DO FEMALES OF *P. CINEREUS* PRODUCE TERRITORIAL PHEROMONES?

Horne and Jaeger (1988) tested the hypothesis that adult females use pheromones, also contained in fecal pellets, to identify female-marked territories, just as Jaeger et al. (1986) had inferred for males. Horne and Jaeger used almost exactly the same design and the four experimental conditions already reviewed from Jaeger et al., except they used noncourting (nongravid) females in place of males. They observed that females NT their own and conspecifics' pellets about equally (unlike males), but females spent significantly more time in both threat and submissive behavior (Fig. 3.1) toward the conspecifics' pellets and more time in their own marked burrows.

Horne and Jaeger (1988) inferred that females of *P. cinereus* (1) mark their own territories with pheromones, (2) prefer their own marked burrows, and (3) either threaten or act submissively toward areas marked by other females. They also conjectured that such pheromones reside in or on fecal pellets, a view later questioned by Simons et al. (1993, 1994). Overall, both males and females acted as if pheromones were important in identifying and communicating one's territorial ownership to self and to others. Surprisingly, though, the behavioral responses of females toward pellets of other females were *more* aggressive than those of males (in Jaeger et al., 1986) toward pellets of other males. This surprising result led to future female versus female and female versus male studies reviewed in sections 7.4 and 7.5, respectively.

5.4 WHERE ARE THOSE PHEROMONES PRODUCED IN MALES AND FEMALES?

Several glandular or nonglandular areas of salamanders in general, and *P. cinereus* in particular, were likely sources of territorial pheromones: (1) the entire surface of salamanders is covered with epidermal/dermal serous glands; (2) fecal pellets from *P. cinereus* and other plethodontid species are impressively large and may contain digestive residues that act as pheromones; (3) earlier morphological studies by Sever (1978b) described the cloacal glands in males of *P. cinereus*, and secretions from these glands could possibly act as pheromones if secreted directly onto the substrate (by cloacal tapping) or if secreted onto the surface of fecal pellets; (4) urinary fluids exiting through the cloaca might be a pheromonal source either alone or if infused into or onto fecal pellets; and (5) specialized glands on the chin of male plethodontid salamanders (Sever, 1976) and middorsal tail base (Sever, 1989, at least in the *Eurycea bislineata* complex) appear to act in the production of courtship pheromones by males, and so they might double as territorial pheromones as well. The former are the mental hedonic gland clusters and the later are the caudal hedonic glands (Sever, 1976, 1989).

Simons and Felgenhauer (1992) tackled the sources of pheromones directly. They tested each of the previously mentioned sources as potential sites of territorial pheromones in behavioral trials with just adult males of *P. cinereus*. They rubbed swabs (1) over male serous glands located on the midventral surface of the tail base and (2) on the shoulder of salamanders; over (3) fecal pellets and (4) fecal material from the colon and (5) products from the urinary collecting ducts; (6) from the mental hedonic glands on the chin and (7) the presumed (for *P. cinereus*) caudal hedonic glands on the middorsal tail base; and (8) a blank control (water) swab. Each male of *P. cinereus* ($N = 39$) was then tested randomly for behavioral responses in circular chambers in each of two conditions: one control and the other experimental. In the control, eight swabs moistened only with dechlorinated water were placed equidistant around the periphery of the circular chamber. In the experimental condition, eight swabs, one from each of the potential pheromonal plus control sources listed previously, were placed randomly and equidistant around the chamber's periphery.

Each male's behavioral responses were compared statistically between the time spent touching or near each experimental swab compared to time touching or near each swab's control (by exactly the same position within the control chamber). The authors also produced scanning electron micrographs (SEM) of the glands of the (1) middorsal shoulder (serous glands), (2) caudal hedonic glands, (3) midventral tail base (serous glands), (4) mental hedonic glands from the chin, (5) a whole body cross-section, and (6) urinary collecting ducts. These six SEM reproductions (*a* through *f* in their Fig. 1) demonstrated that glands were present in each area of study (except for fecal pellets and the cloacal glands, with the latter shown in SEM by Sever, 1978a, 1978b).

The behavioral responses by males to experimental versus control conditions were significantly different. Swabs rubbed over the midventral surface of the tail base and the shoulder (both serous glands) caused exploratory behavior (e.g., NT), as did fecal pellets, fecal material from the colon, and products from the urinary collecting ducts. Swabs from the chin and middorsal tail base (both of which have been shown to produce courtship pheromones) elicited no significant responses.

Simons and Felgenhauer (1992) inferred that males of *P. cinereus* have several secretory (or other) sources that produce pheromonal (or odor) signals used in intraspecific, intermale, social communication. However, their results did not differentiate between pheromones involved in social behavior in general (see chapter 7) and the more pressing problem here of the source or sources of pheromones used in territorial communication.

5.5 FOCUSING ON THE POSTCLOACAL GLAND

Simons et al. (1993) suggested that the postcloacal gland of *P. cinereus* from MLBS should be studied further (at the University of Louisiana at Lafayette) as a source of scent-mark (perhaps territorial) production, because Simons and Felgenhauer (1992) had observed salamanders pressing this gland against previously marked substrates. They termed this marking behavior the postcloacal press (PCP), and they suggested that this gland may be used to "remark" pheromones deposited by other salamanders. The postcloacal gland is an integumental acinar complex found immediately posterior to the cloaca.

Simons et al. (1993) devised a noninvasive technique to occlude (or not occlude) glands with a bilayer of acrylate tissue cement plus surgical rubber cement. Their tests with gland-occluded *P. cinereus* ($N = 26$) versus controls (total of 108 trials) indicated that the bilayer of cement was effective in blocking the deposits of secretory products by the postcloacal gland without hindering the PCP behavior.

This technique was then used in experiments by Simons et al. (1994). They tested the hypotheses that (1) the postcloacal gland produces pheromones used in territorial marking by males of *P. cinereus* and (2) these pheromones induce avoidance by territorial intruders (an inference advocated by Jaeger et al., 1982, in section 3.6). Females were not tested. We only briefly present their methods and results, because we have belabored similar methodologies previously.

Experiment 1 tested the behavioral reactions of intruders ($N = 26$) to substrates either marked or not marked by previous residents. In some replicates, the previous residents had blocked postcloacal glands (blocked by "cement"); in other replicates, the granular glands on the thoracic region had been blocked (the sham control); and in yet other replicates, the substrates were unmarked; that is, residents had not been allowed on those substrates (the general control). The results were surprising relative to the findings of Jaeger et al. (1982). Intruders spent significantly more time on substrates that *had been marked* by unblocked postcloacal glands *and* by unblocked thoracic granular glands compared to the unmarked controls. These results did not support the hypotheses that territorial pheromones alone deflect intrusions by intruders (or at least by same-sex intruders). Instead, Simons et al. (1994) inferred that territorial scent marks do not cause avoidance by conspecifics, which agrees with Gosling's (e.g., 1990) evidence for antelopes and some other mammals: that is, territorial scent marks (pheromones) may function not to repel intruders but instead to reduce territorial defense by establishing the odor identity of the actual resource-holder to the intruders. Whether Gosling's model applies to *P. cinereus* was tested more precisely by Jaeger and Gabor (1993, in section 5.6) and by Simons et al. (1997, in section 5.8).

Experiment 2 by Simons et al. (1994) allowed behavioral interactions between residents and intruders when intruders were introduced to those residents' territorial chambers. Residents were previously blocked so that their substrates could be marked only with scents from either their postcloacal glands or their thoracic granular glands. The results were that residents spent significantly more time in all trunk raised (ATR; Fig. 3.1) toward intruders in trials where they had previously marked the substrates with scents from their postcloacal glands compared to trials in which they had previously marked with scents from their thoracic granular glands. This supported the first hypothesis that the postcloacal gland is a major source of territorial pheromones, because males displayed more aggressively "in defense" on this type of scent-marked substrate.

Intruders initiated contact with residents (touching their bodies) for significantly longer periods of time in both types of trials than residents did with the intruders. This again suggests that territorial scent marks (from either postcloacal or granular glands) do not repel intruders, as in Gosling's (1990) model for some mammals. The major inference here was that, at least for males of *P. cinereus*, the postcloacal gland is a major source of pheromones involved in territorial communication.

Simons et al. (1995) next examined whether juveniles and females of *P. cinereus* have active postcloacal glands. If they do, then both sexes of adults and subadult juveniles might attempt to establish territories. They examined the area of the postcloacal gland in three age groups: (1) neonates (7–11 mm snout-to-vent length [SVL], about two months old, sexes undetermined), (2) a second-year juvenile female (34 mm SVL, about 22 months of age), and (3) an adult male (42 mm SVL, >3 years old). Using SEM images, they determined that all age classes (including the female) had "active" acinar glands in the region of the postcloacal gland, except for the smaller (7 mm SVL) neonate. The glands were deemed to be active because they contained secretory products in their lumina. The inferences drawn were that females also have active postcloacal glands and that this gland becomes active (perhaps involved in social communication) shortly after neonates hatch from their eggs. One hypothesis was that the neonates' secretions from this gland are somehow involved in social communication with their nearby mothers; this hypothesis was later tested by Gibbons et al. (2003; see section 7.18).

5.6 WHAT INFORMATION DOES THE POSTCLOACAL GLAND COMMUNICATE?

Jaeger and Gabor (1993) further tested the hypothesis that the postcloacal gland in *P. cinereus* conveys information to an intruder (and perhaps to territorial neighbors) about the resident's ownership of a territory. Previous observations by Jaeger and Gergits (1979; in section 2.2) had shown that individuals of *P. cinereus* frequently NT each other during territorial conflicts, which implies that NT may be used to gain olfactory information concerning the bodily pheromonal compositions of each other. Jaeger and Gabor (1993) therefore postulated that if the postcloacal gland is involved in social communication (including territorial ownership), then NT should be preferentially directed toward the area of an opponent's body containing the postcloacal gland.

Adult males from Hawksbill Mountain, Shenandoah National Park., were paired as tail-intact territorial residents and intruders that differed by no more than 2 mm SVL. Jaeger and Gabor (1993) recorded, during territorial contests in chambers, the number of contacts and NT by both residents and intruders onto each other and the number of bites. For statistical reasons, they visually divided the salamanders' bodies into five equal length areas: (1) from snout to posterior articulation of the front legs, (2) progressively to the mid-abdomen, (3) to the cloaca, (4) the proximal half of the tail (where the postcloacal gland occurs), and (5) to the distal tip of the tail. During territorial conflicts by each pair ($N = 51$), they recorded (for each of the five areas) the number of (1) NT before bites, (2) bites before NT, (3) bites but no NT, (4) NT but no bites, and (5) no NT and no bites.

Both residents and intruders significantly aimed their NT preferentially toward the portion of the opponent's body containing the postcloacal gland. This provided support for the assertion by Simons et al. (1994) that the postcloacal gland is involved in territorial advertisement. Also, NT between residents and intruders early in a contest (each contest lasted 30 minutes) lessened the likelihood of later bites, especially by intruders. Of importance to the "territorial advertisement hypothesis," residents were seven times more likely than intruders to launch a lightning attack (bite before contact or NT).

Why did residents bother to NT intruders in the area of the postcloacal gland? The territorial advertisement hypothesis would not predict this even though it does predict that intruders would examine that glandular area of residents (especially if the model in Gosling, 1990, is correct). Jaeger and Gabor (1993) speculated that the postcloacal gland provides additional (besides territorial) information concerning identification of territorial neighbors, the sex or kinship of intruders, and/or the fighting abilities (resource holding potential [RHP]) of intruders (*sensu* Maynard Smith & Parker, 1976, concerning asymmetric contests). These topics are further broached in chapter 7, but clearly Jaeger and Gabor's data challenged the territorial advertisement hypothesis for the postcloacal gland as viewed by Simons et al. (1994).

5.7 WHAT SIGNALS DO PHEROMONES COMMUNICATE?

Mathis (1990b) found that both adult males and females of *P. cinereus* from MLBS gain information about the sex and body size of conspecifics through chemical signals (pheromones). She first considered the role of fecal pellets in social communication. Males ($N = 60$) and females ($N = 60$) were fed *ad libitum* with *D. virilis* for two weeks so that each salamander would have a full digestive tract at all times. Then the salamanders were randomly assigned to one of six conditions: males exposed to substrates of C_1: self, C_2: another male, or C_3: a female; and females exposed to substrates of C_4: self, C_5: another female, or C_6: a male. This experiment was conducted blind. The 120 chambers were examined every 2 hours for 9 days to collect the fecal pellets, and passage time through the digestive tract and the final volume of the fecal pellet were recorded.

Females produced pellets significantly fastest when exposed to their own (self) pheromones, while males produced pellets significantly fastest when exposed to pheromones of females. Therefore, Mathis (1990b) inferred that the primary pheromonal function of fecal pellets for females is advertisement of their own "areas" (territories) while for males the primary importance is attraction of mates. (This experiment was performed during the end of the courtship season at MLBS in May.)

Mathis (1990b) then paired future residents and intruders that were either matched for size or of different sizes. Residents and intruders were allowed to establish (mark) territorial substrates in separate chambers for a month, then the resident was removed from its territory and the paired intruder was placed onto that marked territory. The behavioral responses of each intruder were recorded as time spent in (1) ATR, (2) FLAT, (3) look toward fecal pellets (section 3.5), and (4) EDGE and (5) the number of NT to the substrate. Both male and female intruders exposed to pheromones of unseen males produced significantly larger fecal pellets when the intruder and resident were the same sizes (no size asymmetry) but not when the resident was female. Male intruders were significantly more aggressive (ATR) when exposed to the pheromones of unseen residents of the same body size (no size asymmetry) and more submissive (FLAT) when exposed to pheromones of unseen residents of larger size (size asymmetry).

Mathis (1990b) inferred that information concerning body size (an indicator of competitive ability or RHP; Maynard Smith & Parker, 1976) is transmitted via chemical signals by *P. cinereus* and therefore can cause changes in the behavior of conspecifics. She was the first to demonstrate that both fecal pellets and pheromones influence size-dependent social behavioral patterns in *P. cinereus* and (as far as we know) for any other species in the three orders of Amphibia.

Mathis and Simons (1994) tested two hypotheses:

H_1: Male residents will respond to swabs of pheromones from the bodies of conspecific males by re-marking those scented swabs with the residents' own pheromones.

H_2: They will respond by exhibiting agonistic displays in the presence of such swabs.

Adult males (N = 26) established territories in separate chambers, and each resident was paired with another male (N = 26) of the same size (≤1 mm differences in SVL) that was in a separate chamber. Each of the paired nonresident males was rubbed with a cotton swab over the area of the postcloacal gland. Each swab was then placed in the center of a paired resident's chamber, and these swabs became the "intruders" into the residents' territories.

Each resident was tested twice in random sequence, with either a scented swab or a control swab (dampened with water), in a blind experiment. The residents' responses were recorded as seconds spent (1) near the swab, (2) exploring the swab with its head, (3) with snout pressed to the top of the chamber (EDGE), (4) in ATR, and (5) in FLAT and as (6) number of CT in the chamber (see section 3.5), a marking behavior with the male's mental hedonic gland, and (7) number of NT (chemoinvestigation). These data were recorded for 30 minutes for each resident.

Residents exhibited significantly more exploratory behavior to conspecific-marked swabs than to control swabs and significantly more NT in the control condition. The presence of pheromonal swabs did not alter the residents' marking behavior (CT) compared to the control. Residents' agonistic behavioral responses were significantly associated with the body size of the resident: larger male residents were more aggressive and smaller residents were more submissive toward pheromonal swabs even though those swabs came from males the same size as the resident. These results did not support H_1 that males attempt to mark over the pheromones of other (intruding) males with their own pheromones. However, H_2 was supported: that just postcloacal pheromones from unseen "intruders" cause increases in threat or submissive behavior by residents.

Most important, though, Mathis and Simons (1994) demonstrated that the absolute body size, not relative body size, of a male territorial resident influences his agonistic behavioral responses toward conspecific males' pheromones. That is, even though the swabs came from intruders that were the same body size as the paired residents, smaller residents responded more submissively and larger residents more aggressively toward those swabs. In retrospect, this difference conforms to expectations of territoriality in that larger males or females of most species are more likely to compete for and defend territories more successfully than are smaller conspecifics. Mathis (1990a, in section 3.7) had demonstrated this with experiments in the forest at MLBS with both sexes of *P. cinereus.*

5.8 SCENT MATCHING AND TAIL AUTOTOMY

Simons et al. (1997) further tested Gosling's (1990) model of the functions of territorial "scent marks" (a term that we use synonymously with "territorial pheromones"). He had proposed this concept for territorial scent-marking mammals, such as antelopes, but Simons et al. (1994) and Jaeger and Gabor (1993) had inferred, from their experiments, that this model may apply to *P. cinereus* as well.

Gosling's (1990) scent-matching hypothesis states that territorial scent marks do not function to repel intruders. Instead, these scent marks announce, to the intruder, the presence of the resident and the resident's "social rank" and ability to defend the territory. Also, by scent matching the scent marks on the ground in the territory with the scents detected on the resident's body, the intruder can identify the true territorial resident from other simultaneous intruders that might be in that territory. In essence, territorial scent marks may ultimately function to decrease the resident's costs of territorial defense. Such costs can be severe for *P. cinereus,* because biting contests can lead to scars in a resident's or intruder's nasolabial grooves, which reduces that salamander's ability to detect ambient olfactory information, such as the locations of prey, mates, or predators (Jaeger, 1981b, in section 3.5).

Simons et al. (1997) used only males of *P. cinereus* to test under two conditions whether intruders would enter a specific part of a test chamber: C_1: in experimental trials, containing the marked boundary of the resident, and C_2: in control trials, in a substrate previously marked by a nonresident conspecific male (the unseen "source" animal). These two conditions were intended to differentiate the behavioral responses of intruders toward C_1 (the real territorial resident) from C_2 (just another unseen intruder in or near a resident's territory).

Each resident had been fed 40 *D. virilis* in 72 hours so that it would have a territory worth defending against intruders. Each resident's ($N = 29$) chamber contained three contiguous sections, labeled 1, 2, and 3. In experimental trials (C_1), the resident had marked sections 1 and 2 and a source animal ($N = 29$) had marked section 3. In control trials (C_2), the resident had marked only section 1 while the source animal had marked sections 2 and 3. During this set-up period (72 hours), the resident and source animal were separated by an opaque barrier. After the set-up period, the source animal was removed and the opaque barrier was moved to between sections 2 and 3; a mesh screen was placed between sections 1 and 2, which thus physically, but not visually or olfactorily, separated the resident from the forthcoming intruder ($N = 29$).

The intruder's own chamber was then connected to the resident's chamber by a tunnel such that the intruder could choose to enter only section 2 of the resident's chamber. Once in section 2, the intruder was on a substrate marked either by the resident (in experimental C_1) or by the source animal (in control C_2). Section 3 always contained the markings of just the source animal. To avoid asymmetric contests, each resident, its source animal, and its intruder were matched for size (< 2 mm differences in SVL), and each trio was used twice, in random order, for the experimental and control conditions.

During each 30-minute trial, Simons et al. (1997) recorded (1) the time spent in ATR for the resident and intruder, (2) the time spent in FLAT for the resident and intruder, (3) time spent in section 2 for the intruder, (4) the number of NT (sampling for scent marks) and (5) the number of gular pumps (presumed to sample for airborne olfactory cues: see section 5.10).

Simons et al. (1997) inferred that *P. cinereus* conformed to three major hypotheses subsumed within Gosling's (1990) scent-matching model.

> H_1: Intruders should not be deterred by the scent marks of territorial residents.

For *P. cinereus*, intruders did not differ significantly ($p = 0.36$) in times spent in section 2 between experimental and control conditions. That is, intruders approached residents despite the residents' scent marks in section 2 during the experimental trials, which fit well with Gosling's hypothesis.

> H_2: Territorial pheromones should "intimidate intruders" or "enhance the confidence of the residents" during territorial contests (*sensu* Gosling, 1990).

For *P. cinereus*, intruders hardly ever resorted to FLAT in either experimental or control conditions, even though the resident could be seen and smelled in both conditions. Therefore, there was no evidence for "intimidated" intruders. However, residents did show "enhanced confidence," because they spent significantly ($p = 0.02$) more time in ATR during experimental trials (when intruders were on the residents' marked substrates in section 2) than during control trials (when intruders were on substrates marked not by the residents but marked by the source animals). This was inferred to mean that residents' had enhanced confidence on their own scent marks.

> H_3: Territorial scent marks should provide intruders with information about the "status" of the resident, which can be used by intruders and residents in their decisions concerning subsequent aggressive and submissive interactions.

(Note that Mathis, 1990b, had shown that substrate scent marks by *P. cinereus* convey status information to conspecifics concerning the body size of the salamander leaving those scent marks, and Mathis and Simons, 1994, had shown that males of smaller body size are less aggressive toward scent marks of same-size conspecifics than are males of larger body size.) This hypothesis was difficult to evaluate in this experiment, because each intruder in the experimental condition was physically face-to-face with the resident and with his pheromones on the substrate and perhaps with volatile pheromones diffusing in the air from the resident's body. Which of these influenced the intruder's behavior was unknown. Still, two results might support this hypothesis. One was that residents spent significantly ($p < 0.01$) more time NT in the experimental condition than in the control. This suggests that residents tried to gain more olfactory information about the intruders when the intruders were on the residents' marked substrates. The second, more indicative result was that intruders displayed significantly ($p < 0.05$) more gular pumps in the experimental condition, as if trying to gain more olfactory information from volatile pheromones coming from the residents' bodies on the other side of the screen. Section 5.10 summarizes three experiments concerning volatile pheromones from *P. cinereus*.

Overall, these results indicate that territorial (and other social) pheromones convey a wealth of information to intruders concerning a resident's status and ownership of the area. They also add credence to the notion that the territorial scent marks of *P. cinereus* conform to the hypotheses of Gosling's (1990) scent-matching model, a model that had been derived from his observations of African antelopes.

Wise et al. (2004) studied the influence of tail autotomy on the territorial markings and behavior of *P. cinereus*. Territorial salamanders and lizards, of some species, can autotomize their own tails when bitten by a predator (which may then chase the wiggling tail instead of the rest of the animal's body) or when bitten during territorial contests (in *P. cinereus*, the winner usually eats the loser's fat-filled tail). Loss of the tail can cause an immediate reduction of the loser's RHP or "status" (*sensu* Gosling's, 1990, scent-matching model). That is, tail loss causes an increase in the loser's asymmetry with contenders, because it becomes immediately shorter in body length. Wise et al. examined (1) the marking behavior of tailed and tailless residents, and (2) the behavioral responses of intruders to scent marks produced by tailed and tailless residents. Mathis (1990b) had shown that information about an individual's body size (and its RHP) can be conveyed by its scent marks. The two experiments by Wise et al. were expected to shed light on whether tail autotomy decreases the loser's RHP (commonly seen in the forest at MLBS; section 3.10).

Salamanders were collected and those within a tetrad were matched for size (≤ 1 mm SVL) to reduce the asymmetries between future residents and intruders ($N = 18$ tetrads, a total of 72 males). Salamanders were divided as follows: randomly chosen future tailed and tailless residents, and tailed and tailless intruders. Tail autotomy was induced by pinching (with forceps) a given section of the tail such that the salamander would detach 85% of his original tail length, well posterior to the postcloacal gland. *Plethodon cinereus* has a specialized way of detaching its tail at the place near where it had been bitten or, in this case, pinched.

Experiment 1 tested the effect of tail condition on scent marking. Tailed and tailless residents, fed *D. virilis ad libitum*, established separate territories in individual chambers. As they did so, the number of PCP were recorded during each 30-minute test. The hypothesis was that if tail loss reduces the RHP of residents (i.e., ability to resist intruders), then they should differ in marking behavior between experimental (tailless) and control (tailed) conditions. Wise et al. (2004) found that tailless salamanders marked, by PCP, significantly more per 30 minutes than did tailed residents. This may imply that tailless residents "tried to overwhelm" future intruders with a superabundance of pheromones on their substrates.

Experiment 2 tested the responses of tailed and tailless intruders to substrates previously marked by (now unseen) tailed and tailless residents. This was a blind experiment. All residents were removed from their chambers after 5 days of marking, and a tailed or tailless intruder was introduced to the chamber of a tailed or tailless resident. Wise et al. (2004) recorded, for each 30-minute test, the time that each intruder spent in the ATR and FLAT postures. Each intruder was given five *D. virilis* after the first 15 minutes so that he would find the territory rich in prey and, thus, "worth an attempt to take over" that territory. The null hypothesis was that intruders could not distinguish the tail condition of the unseen residents merely by their scent marks.

Wise et al. (2004) found that tailed, but not tailless, intruders exhibited significantly less ATR in chambers previously marked by tailed residents than in chambers previously marked by tailless residents. This implies that scent marks, in the absence of visual cues, provide male intruders with information about the RHP of male residents. However, the increased marking by tailless residents may provide a benefit to them, because tailed intruders exhibited significantly less aggression on substrates where tailless residents had performed more PCPs.

The overall impressions from these two experiments are that tail-autotomized residents compensate for tail (and RHP) loss by abundantly marking their territories, and this does reduce aggression from tailed intruders. However, intruders can still otherwise estimate the tail condition (and thus RHP) of residents just by their scent marks, presumably from the postcloacal gland.

Territorial scent marks may be of major importance to the RHP of *P. cinereus* at MLBS. The density of this species there was 2.8 salamanders/m^2 on the forest's floor (Mathis, 1990a), and recent tail autotomy was commonly observed by Wise and Jaeger (2016) and by Gillette (2003) who marked–recaptured 3,487 individuals in her three years at MLBS (see section 7.19). Therefore, territorial intrusions and combats are probably common at MLBS due to such high densities. Few predators on adults of *P. cinereus* occur at MLBS, except for shrews and large coleopterans (based on our observations), so territorial conflicts perhaps account for a large percentage of tail autonomies there, but see contrary evidence in Wise and Jaeger (2016) in section 3.10.

5.9 DO TERRITORIAL PHEROMONES AID IN HOMING BEHAVIOR BY *P. CINEREUS*?

Homing behavior has been well studied in the genus *Plethodon* and occurs when a salamander displaced from under a rock or log returns to that same site. Kleeberger and Werner (1982) previously reported that *P. cinereus* had returned within 24 hours with 90% accuracy after being displaced 30 m and with 25% accuracy from 90 m displacements. Also, return routes were generally in a straight line. This is a remarkable achievement for an adult salamander that is only about 9 cm long, has the thickness of a pencil, and must walk through leaf litter to reach home! Homing behavior would be important for *P. cinereus* so as to return to its "home" territory after foraging trips into the leaf litter or after climbing plants during wet nights (see section 3.2). Quick returns may be due to a territorial salamander's need to return before an intruder tries to set up its own territory there (by scent marks and displays) in the absence of the original owner.

In the forest at MLBS, Jaeger et al. (1993) tested the hypothesis that *P. cinereus* uses territorial pheromones to relocate a home site. They located 150 adult males and females that were under rocks (presumably their territories) and divided them into three groups. Each salamander was uniquely toe-clipped for later identification, but females and males were treated as a single dataset, because Kleeberger and Werner (1982) had found no significant differences between the sexes in abilities to return to home sites.

In group 1 (N = 55 salamanders), Jaeger et al. (1993) merely lifted then returned each rock to its original site, leaving the dirt underneath undisturbed. The salamander found under that rock was lifted and then placed next to its rock. Group 1 was the control condition to determine what percentage of recaptures could be expected for salamanders that had been handled but not displaced away from their home rocks.

In group 2 (N = 43), they returned the rock to its original site after disturbing (with a trowel) the soil underneath it (as a control for group 3). The salamander was lifted and displaced 5 m away from its home rock in a random cardinal direction. They chose 5 m because *P. cinereus* can easily return home from that distance.

In group 3 (N = 52), they moved the home rock and underlying soil 5 m away in a random direction and then transplanted another same-size rock from >5 m away to the site of the original home rock. New soil (not from under a rock) was used to fill the cavity under the newly transplanted rock. The home rock's salamander was lifted and moved to a place 5 m equidistant from both the now-moved home rock and the newly transplanted non-home rock. The assumption was that the rock and/or soil under it would contain pheromones needed for homing in groups 2 and 3.

Based on the hypothesis that territorial pheromones are cues for homing, Jaeger et al. (1993) predicted that salamanders in all three groups would return to the home rock, even though in group 3 the home rock and soil had been moved. The results did not provide support for the hypothesis or the prediction. In group 3, all salamanders returned to the original sites of capture, not to the home rocks and soil that had been at those sites. Also in group 2, the displaced salamanders returned to their home sites and rocks, and none of the statistics for groups 2 and 3 were significantly different from returns by nondisplaced salamanders in group 1.

Jaeger et al. (1993) proposed four a posteriori alternative hypotheses to explain their results.

H_1: *Plethodon cinereus* does not use odor cues for homing. They rejected this hypothesis because a previous study had shown that *P. jordani* homes by chemical cues alone (Madison, 1969).

H_2: Other chemical information may be used while homing, such as odors of plants, fungi, and decaying vegetation surrounding the home territory.

H_3: Salamanders may respond to their own territorial pheromones if those pheromones are distributed over a wider area than just on or under their home rocks.

H_4: Salamanders may form a cognitive map of territorial pheromones deposited by other salamanders in the homing animal's area of the forest.

This last hypothesis is particularly attractive, because *P. cinereus* does not reach sexual maturity until 2 to 3 years of age, and after that it does not grow to maximum size (and RHP) needed to gain and defend a territory for many years. Therefore, these salamanders may spend perhaps 8 to 10 years before establishing territories, during which time they behave as nonterritorial "floaters" in the population (Gillette, 2003, in section 7.19). This time could be used to form a cognitive map of who lives where before a salamander establishes its own territory amid those surrounding conspecific pheromones. Section 9.5 reviews how *P. cinereus* employs individual recognition memory to detect the pheromones of conspecifics. Still, H_2, H_3, and H_4 are ripe for future experiments in forests where *P. cinereus* is abundant.

5.10 ARE PHEROMONES VOLATILE?

Based on previous territorial research with *P. cinereus* (in chapters 3 and 4), it had been assumed that territorial pheromones do not diffuse easily into the air. The assumption was that such pheromones are composed of large, heavy molecules that, by not diffusing quickly, would provide reliable, long-term scent marks defining a particular salamander's territory and its boundary. Nose tapping to the substrate by invaders and residents had been frequently observed in *P. cinereus* suggesting that pheromones are located on the substrate and detected by the nasolabial cirri and sent up the nasolabial grooves into the nares. In a territorial experiment, however, Simons et al. (1997, in section 5.8) recorded gular pumping, in which a salamander rapidly pulses its throat (with its mouth shut) such that airborne odors would flow into the nares. Airborne odors would be useful to residents and intruders in order to gain information (e.g., RHP) concerning their opponents at a distance beyond which biting attacks could occur and cause injury. Therefore, a research program was begun to learn whether or not territorial and other pheromones of *P. cinereus* are volatile.

Martin et al. (2005) designed an experiment with four experimental treatments (T_1–T_4) and a control. Each treatment contained two objects in a chamber: one inside a perforated Petri dish and the other outside the dish. In the control, the two objects were blanks (dampened, rolled filter papers). In T_1, the rolled filter paper in the dish contained the scent marks (over 5 days) of a source male and the rolled paper outside the dish was a blank. In T_2 a blank was inside the dish and the source male's marked paper was outside the dish. In T_3 a source female's marked paper was inside the dish and a blank was outside. In T_4 a blank was inside the dish and a source female's marked paper was outside.

Before this set-up, a resident male ($N = 30$) had established a scent-marked territory in each chamber and then was removed to allow the placement of the two objects. The test began when the resident was returned to his chamber and was confronted with the two new objects in its territory. Each test lasted 15 minutes during which Martin et al. (2005) recorded each resident's time spent in ATR and the number of NT to both the perforated dish and the filter paper outside the dish. Each resident was tested five times in random order, once with the control and once with each of T_1 to T_4. As with other experiments reviewed here, the same observer watched a given salamander through all five tests to reduce interobserver biases when recording data.

The results were unambiguous. Male residents were significantly more aggressive (ATR) toward male pheromones both inside and outside the dishes compared to the control ($p < 0.0001$), suggesting that males could detect volatile male scent marks coming from inside the perforated dishes. The results from the experiment did not suggest that males could differentiate between volatile pheromones of females and males, but males were significantly more aggressive toward male scent marks outside the perforated dish than toward such female scent marks, where NT were used to sample directly those scent marks. Residents also NT dishes containing both male and female scent marks significantly more frequently compared to the blank control. The authors inferred that males need not patrol their territories in search of male intruders but can detect such intruders at a distance via their volatile pheromones. They also suggested that males may need direct contact (by NT) to distinguish between male and female intruders. This hypothesis was tested next by Dantzer and Jaeger (2007a).

Dantzer and Jaeger (2007a) tested the null hypotheses that neither territorial males nor females of *P. cinereus* can distinguish between volatile odor cues from source (pheromone donors) males and females. Resident males ($N = 29$) and nongravid females ($N = 30$) established territories in individual chambers and then were confronted with each of six treatments, in random order, using source male pheromones, source nongravid female pheromones, and a blank control. The source and residential salamanders had been collected from different forested sites separated by ≥1 km so that residents would not have had prior experiences with the source animals' pheromones. The pheromones and blank control were on rolled filter papers placed in plastic tubes with the ends covered by a screen (allowing possible volatile pheromones to diffuse into the chambers). These tubes contained (1) source male's scent marks deposited over 1 day or (2) over 5 days, or (3) 1-day control blank; and (4) source nongravid female scents deposited over 1 day, or (5) over 5 days, or (6) a 5-day control. Sex and reproductive condition were determined by the method described in section 3.4. A test began when a territorial male or nongravid female found, in random sequence, one of the six tubes in its territory. During each 15-minute test, Dantzer and Jaeger (2007a) recorded the resident's amounts of time in (1) ATR and (2) touching the tube, and the number of NT to (3) the substrate and to (4) the tube. The statistical tests, among pheromonal treatments and between these treatments and the appropriate 1-day or 5-day control, were complex but can be summarized briefly.

Both male (174 tests total) and female (180 tests) residents spent significantly more time in ATR when exposed to volatile conspecific chemical signals (diffusing from the tubes) from same-sex source salamanders than from opposite-sex source animals. This suggests that both males and females of *P. cinereus* can distinguish the sexes of (unseen) intruders based just on their volatile pheromones; however, this was significant only for 5-day source pheromones. The results for 1-day source pheromones were ambiguous. Therefore, substrates may need to be scent marked for >1 day but <5 days before salamanders can distinguish between the sexes based on volatile pheromones. A similar statistically significant pattern was found for number of NT to the substrate but not to the tubes.

Dantzer and Jaeger (2007a) inferred that volatile pheromonal signals do convey sexual information to males and females and that such airborne pheromones may also influence the spatial organization of male and female territories on the forest floor, as seen by Wicknick (1995) in section 6.7. This hypothesis could lead to future research in forests where *P. cinereus* is abundant and, thus, where territories may be closely spaced. Dantzer and Jaeger inferred that both males and nongravid females can distinguish the sexes of conspecifics by their volatile pheromones, but a male should be more prone to search for a gravid female (with yolked ova and ready for sperm) instead of a nongravid female (still recovering from brooding her clutch of eggs). Therefore, Dantzer and Jaeger (2007b) tested the hypothesis that males can distinguish between gravid and nongravid females' volatile pheromones.

Dantzer and Jaeger (2007b) used the same experimental design as in Dantzer and Jaeger (2007a) with the following exceptions. Only males of *P. cinereus* were used as territorial residents. Each male ($N = 30$), in random order, was exposed to three treatments: a tube containing a rolled filter paper marked for

5 days by (1) a source gravid female, (2) a source nongravid female, or (3) water as a control. Randomization statistics (Adams & Anthony, 1996) were used for this experiment to compare the responses of residents to both types of female source animals and between them and the control. Randomization techniques are thought to yield more statistical power than some nonparametric and parametric statistics (see Gillette, Jaeger, et al., 2000, and Dantzer & Jaeger, 2007b, for rationales and statistical references).

Dantzer and Jaeger (2007b) found that male residents exhibited significantly more ATR toward chemical signals from nongravid females than toward gravid females ($p = 0.010$) and the control ($p = 0.0001$). Nose taps to the substrate yielded no significant differences. Therefore, Dantzer and Jaeger inferred that territorial males can distinguish the reproductive state of a female by volatile pheromones alone. Increased threat behavior toward nongravid females may be a male's attempt to drive her away from his territory; she would not be willing to court but she may eat the often scarce prey in his territory. However, see a conflicting "ESS dating game" by Hom et al. (1997) in Part section 7.6. On rainy nights, wet observers at MLBS have often seen one or more males following a female up a tree, so pheromonal tracking of gravid females probably occurs outside as well as within territories.

This concluded a long research program (1986–2007) to determine if *P. cinereus* produces territorial pheromones or scent marks (it does) and where they come from (several glandular sources) and their informational signals to conspecifics (many). More pheromonal experiments are reviewed here, but they concern tests of particular hypotheses outside the general scope of the research reported in this section.

Many other hypotheses need yet to be tested experimentally concerning pheromonal communication in *P. cinereus* and other salamanders. One hypothesis is that males can differentiate between already inseminated versus uninseminated courting females and thus avoid sperm competition and cryptic female choice (Eberhard, 1996). Studies of social monogamy, mate guarding, and sexual coercion by *P. cinereus* (in chapter 7) and the cognitive abilities of males to remember familiar conspecifics (in chapter 9) complicate an overall assessment of pheromonal communication in *P. cinereus.*

5.11 SELECTED RECENT RESEARCH BY OTHERS: PHEROMONAL COMMUNICATION

Following the work of Martin et al. (2005), many studies have shown that plethodontid salamanders detect pheromones via the vomeronasal olfactory system and that these salamanders increase NT when the odor of a conspecific (and others) is found on the substrate in question. For example, Schubert et al. (2008) examined sex differences in chemoinvestigation in the red-legged salamander, *Plethodon shermani*. They found that males NT neutral substrates and those moistened with female body rinses more than did females. However, males and females did not differ in responsiveness of vomeronasal organ chemosensory neurons to male mental gland extract or female skin secretions, indicating that males are not more responsive to all chemosensory cues. Similarly, Dalton and Mathis (2014) found that male and female Ozark zigzag salamanders, *Plethodon angusticlavius*, can discriminate between chemical substrates marked by male and female conspecifics. Further, males NT more on own-marked substrates than did females.

In an attempt to tease out multimodal cue use, Duhaime-Ross et al. (2013) examined the response of adult and juvenile *P. cinereus* to visual or chemical cues of conspecifics. They found that adults of *P. cinereus* increased NT to the substrate in response to chemical cues but not visual cues from conspecifics. In contrast, juveniles increased NT in response to either visual or chemical cues of conspecifics. These results support the finding that adult salamanders primarily rely on chemical cues for communication about conspecifics. Juveniles, on the other hand, NT at a low rate, and when conspecific visual or chemical cues are present they increase this behavior. One hypothesis for this finding is that learning is involved in the process of detecting conspecifics in *P. cinereus* (see section 9.5).

Jaeger et al. (1986) proposed that more information was needed about what signals are transmitted to receivers by pheromones. Also using *P. cinereus*, Chouinard (2012) followed Jaeger's proposed research direction by exploring the type of information that is communicated in chemical signals based on diet, how fast diet changes, and if these cues are condition dependent. The high-quality diet consisted of 0.1 g of white worms (*Enchytraeus albidus*), and the low-quality diet consisted of 12 to 14 wingless fruit flies (*D. melanogaster*). He found that gravid females used chemical cues to detect changes in the diet quality of male pairs within a week of altered feeding. Females NT males with high-quality diets more than those with low-quality diets. Additionally, Chouinard found that unlike the results in Walls et al. (1989), fecal pellets were not necessary for detecting diet quality, because NT by females in response to chemical cues of males changed over short feeding periods. Chouinard used protein assays of the mental and postcloacal glands to explore how these changes were detected so quickly. He found that males fed high-quality food had significantly more proteins present in both signaling glands than males fed low-quality food. These results further support the hypothesis that the secretory products in the glands are involved in communication of territoriality and mate quality. Additionally, males provide pheromonal honest signals of mate quality. A more in-depth review article by Woodley (2015) explored the types of chemosignals that have been identified and how hormones modulate these signals in relation to amphibian reproduction.

The chemical signals that individuals encounter contain many types of information about conspecifics, including information about diet. In chapter 6 we explore communication between species during interspecific conflicts.

6

Interspecific territoriality and other interspecific behavioral interactions

Jaeger's research group has long held to the paradigm that research results from behavioral experiments conducted in the laboratory are merely suggestive of intraspecific and interspecific behavioral interactions. Such suggestive results then need to be experimentally tested in the species' natural habitats (when possible) to be more definitive or conclusive. For example, laboratory experiments testing for intraspecific territoriality in *P. cinereus* (in chapter 3) were conducted in small, sterile chambers with no predators or environmental perturbations. When these salamanders were given a limited food supply, usually not to satiation, they may have been "forced" to act aggressively with opponents. Such behavior may not actually occur in the natural forested habitat where salamanders (not in confined spaces) may be more concerned with avoiding predators and dealing with mates, weather perturbations, diseases, and parasites than with interference competition. Therefore, the laboratory experiments reported in chapter 3 merely suggest territorial behavior within *P. cinereus*. Not until Mathis (1990a, in section 3.7) conducted her competitive (territorial) release experiment in the forest at Mountain Lake Biological Station (MLBS) did the hypothesis for intraspecific territoriality gain more definitive support (at least for *P. cinereus* at MLBS). Jaeger and Halliday (1998) considered the difference between exploratory (e.g., in the laboratory) and confirmatory (e.g., in the natural habitat) research.

The same problem of laboratory research versus research in the natural habitat also applies to tests of interspecific territoriality between *P. cinereus* and *P. shenandoah*. We return to that proposal now.

6.1 INTERSPECIFIC TERRITORIALITY BETWEEN *P. CINEREUS* AND *P. SHENANDOAH*

By 1990 enough information had been gained concerning intraspecific territoriality by *P. cinereus* that Griffis and Jaeger (1998) could return to Jaeger's (1972) hypothesis that *P. cinereus* dominates *P. shenandoah* by interspecific territoriality on Hawksbill Mountain, Shenandoah National Park (SNP), Virginia. They conducted two experiments there. Experiment 1 was a competitive (territorial) release experiment on the forest floor in the area of parapatry between the two species. This was to test the hypothesis that, in the natural habitat, *P. cinereus* does indeed hold territories against *P. shenandoah*, as originally proposed by Jaeger (1972). Experiment 2 was in a laboratory at SNP to test how *P. cinereus* behaviorally gains territorial dominance over *P. shenandoah*: hypothetically by interference, not exploitative, competition as suggested in section 2.2.

Plethodon shenandoah on Hawksbill Mountain forms a metapopulation, with the large talus on top of the mountain housing the "source" population (*sensu* Pulliam, 1988) and at least five, much smaller taluses at lower elevations housing the "sink" populations (Jaeger, 1970, Fig. 2). Each of these is completely surrounded by *P. cinereus* in the deep soil on the forest floor. The largest of the sinks is only ~9 m from the source (Jaeger, 1970, Fig. 4) and thus should be easily available to *P. shenandoah* moving from the source to the sink, and some *P. shenandoah* salamanders were found on the soil between these two taluses (Jaeger, 1970, Fig. 4). The other four smaller sinks are further away from the source and so may be more difficult for *P. shenandoah* to find when moving from source to sinks. Of underlying interest to Griffis and Jaeger (1998) was not only if individuals of *P. cinereus* hold territories against *P. shenandoah* but also if such presumed territoriality by the former inhibits (or even prevents) movements of the latter species across deep soil from the source to the sinks. If this were indeed the case, then Hawksbill Mountain would be a rare example of interspecific territoriality in a metapopulation adding to the extinction-prone status of an endangered species. Note that SNP is a national park, and as such both *P. shenandoah* and *P. cinereus* are completely protected against human disturbances, even though the much-hiked Appalachian Trail crosses both of the salamanders' habitats.

Griffis and Jaeger's (1998) experiment 1 was a classic competitive release experiment, in the area of parapatry within 10 m of the edge of the source talus on Hawksbill Mountain. The authors located 100 rocks that met two criteria: (1) each had under it one adult *P. cinereus*, and (2) each rock covered enough soil (215–940 cm^2) to house multiple salamanders in the absence of competition. They randomly assigned the 100 rocks to control ($N = 50$) and experimental ($N = 50$) conditions. For the controls, each *P. cinereus* was returned under its home rock. For the experimental rocks, each *P. cinereus* was removed and released at Hawksbill Gap, 0.75 km away from the study site to prevent homing behavior (which can occur up to 90 m displacement, according to Kleeberger & Werner, 1982). The soil and any territorial pheromones therein were not disturbed under any rock, and invading red-backed salamanders under experimental rocks were removed to leave those sites empty for *P. shenandoah*. Rocks were turned 3 to 8 days after each rainfall, when salamanders should be leaving the leaf litter due to increasing desiccation stress. This experiment was conducted twice, in 1991 (June–August) and 1992 (May–July), during the noncourtship summers. The statistical results were based on (1) the number of *P. shenandoah* invading control versus experimental rocks and (2) the number of days until such invasions occurred.

During 1991, individuals of *P. shenandoah* invaded significantly more experimental than control rocks, and the same was found in 1992. Latent times to these invasions were significantly shorter for experimental ($N = 40$ days) than for control ($N = 52$ days) rocks in 1991, the only year in which invasion times were monitored. Griffis and Jaeger (1998) inferred that these data supported the hypotheses that (1) *P. cinereus* does successfully defend territories (home rocks) against *P. shenandoah* and (2) such interspecific territoriality leads to the extinction-prone status of the latter species, by inhibiting migrations from the source to the sink populations, or vice versa. Recall from section 2.1 that only those *P. shenandoah* as large as or larger than the largest *P. cinereus* were found >3 m from the source talus (Jaeger, 1972), suggesting that source-to-sink migrations are rare. Also, Jaeger (1980a) had found that one small sink population became extinct due to a prolonged drought and had not been reestablished by *P. shenandoah* over the last nine years of his censuses. This also suggests limited source-to-sink migrations, assumed to be due to aggression from *P. cinereus* in the intervening deep soil habitat.

Griffis and Jaeger (1998) speculated that because *P. shenandoah* cannot easily invade home rocks of *P. cinereus*, perhaps the former species merely thrives as a large population in the leaf litter (on deep soil) between territories of *P. cinereus*. If so, this would provide an easy path of migration from source-to-sink populations. Therefore, they used 1 m^2 quadrats to census, in late summer 1992, the densities of both species in the leaf litter where the two competitive release experiments had been performed. The 50 quadrats were randomly placed 1 to 15 m from the edge of the source talus, and therein the leaf litter was searched and cover objects (rocks and logs) were turned in search for salamanders (excepting those in fossorial retreats of course). The researchers found 60 *P. cinereus* (1.2/m^2) and seven *P. shenandoah* (0.14/m^2). Of the latter species, five were adults, and none was found 11 to 15 m from the edge of the source talus. These data suggest that *P. shenandoah* is rare in areas between (as well as within) territories of *P. cinereus*. The largest of the sink taluses, however, may receive some immigrants from the source talus, which is only ~ 9 m distant from that sink, because six of the seven *P. shenandoah* were found 7 to 11 m from the source talus. Thus interspecific conflicts may inhibit but not prevent source-to-sink migrations in this metapopulation. Griffis and Jaeger also found that, for experimental rocks, the original inhabitants were significantly larger in snout-to-vent length (SVL; $p = 0.01$) than the conspecific invaders that replaced them. This conforms to the results, reviewed in section 3.7, that only very large adults of *P. cinereus* can establish and/or defend territories against conspecifics on the forest floor.

While experiment 1 yielded interesting results, experiment 2 by Griffis and Jaeger (1998) provided mostly frustrating results. This was conducted in a light- and temperature-controlled laboratory at SNP with *P. cinereus* and *P. shenandoah* freshly collected from our research site on Hawksbill Mountain (and released there at the conclusion of this behavioral experiment). The working hypothesis was that Shenandoah salamanders are aggressively inferior compared to *P. cinereus*, based on data from experiment 1 and from previous research on Hawksbill Mountain by Jaeger (1970, 1971a, 1972). Both species were placed individually in separate chambers containing a moist substrate and a cover object (small Petri dish) as a refuge. Tubifex worms were placed under the slightly elevated Petri dish. After a 5-day set-up period, the experiment began with four completely randomized conditions: C_1: a resident *P. cinereus* invaded by a *P. shenandoah* ($N = 29$); C_2: a resident *P. shenandoah* invaded by a conspecific for intraspecific levels of aggression ($N = 29$); C_3: a surrogate replica of a salamander as a resident invaded by *P. shenandoah* (a control for C_1; $N = 30$); and C_4: a resident *P. shenandoah* invaded by *P. cinereus* ($N = 30$). Salamanders were randomly assigned to these four conditions regardless of differences in body sizes (and thus differences in resource holding potential [RHP]) so as to mimic aggressive contests likely to occur in the forest on Hawksbill Mountain. Data were recorded, after the usual 15-minute habituation period, for 15 minutes, as (1) number of bites by each salamander (usually brief, harmless "nips"); (2) total time of biting by each (more prolonged, dangerous "wrestling matches" during which salamanders mutually bite each other while thrashing about on the substrate); (3) the all trunk raised (ATR) threat posture; (4) the FLAT submissive posture; and (5) location within 2 cm of the tubifex worms under the Petri dish.

While both species nipped and wrestled with each other, only 2 of the 20 statistical comparisons among C_1 to C_4 were significant. Also, a similar number of statistical comparisons was made between the sexes of *P. cinereus* and, separately, *P. shenandoah*; only one was significant. Nonsignificant p values ranged from 0.164 to 0.894 while the three significant p values were not far below $p = 0.05$. These results were not impressive with alpha set at 0.05! Griffis and Jaeger (1998) concluded that they could not reject the null hypothesis that the two species are equally aggressive during interspecific encounters. Therefore, if *P. cinereus* does dominate *P. shenandoah* in the forest on Hawksbill Mountain, as in experiment 1, it must use subtler tactics than revealed in experiment 2. One such subtle tactic may be that *P. cinereus* directs its bites toward the nasolabial grooves (NLG) of *P. shenandoah*, thus causing scar tissue to form in those grooves and leading to the recipient eventually becoming olfactorily impaired (a long-term cost of aggression). Red-backed salamanders employ this biting tactic against conspecifics at SNP (Jaeger, 1981, in section 3.5), but Griffis and Jaeger (1998) foolishly failed to record *where* the two species bit each other in experiment 2. A posteriori, microscopic examination of the NLG of *P. shenandoah* was not possible, because one cannot kill members of an endangered species either ethically or legally.

The publication of Griffis and Jaeger (1998) ended all research with *P. shenandoah*, because federal, state, and SNP authorities refused future research permits of any kind. Perhaps those agencies were alarmed that an endangered species was actually bitten by *P. cinereus* in experiment 2! Consequently, this was a sad ending to an unfinished, off-and-on research program that had lasted for 31 years (1967–1998). Next we turn to behavioral research concerning interspecific interactions between *P. cinereus* and other (not endangered) species of Plethodontidae.

6.2 RULES OF ENGAGEMENT WITH JUVENILES OF *P. GLUTINOSUS*

While *Plethodon glutinosus* (the slimy salamander) is rarely encountered on Hawksbill Mountain in the Blue Ridge Mountains of western Virginia, it commonly shares the forest habitat with *P. cinereus* at MLBS, in the Appalachian Mountains of southwestern Virginia (Fig. 6.1A). The two species probably do not compete for prey as adults, because *P. glutinosus* grows to a much larger maximum length (~21 cm total length at MLBS) and obtains greater bulk, whereas *P. cinereus* is smaller in maximum length (~9 cm total length at MLBS) and bulk. However, the juveniles and subadults of the former species must grow through all of the sizes of *P. cinereus* until they exceed ~9 cm total length. Because the two species are spatially mixed on the forest floor, competition for prey may occur between the young of *P. glutinosus* and both juveniles and adults (>35 mm SVL) of *P. cinereus*. Lancaster and Jaeger (1995) tested two hypotheses:

H_1: That juveniles of *P. glutinosus* seldom are found in territories of adult *P. cinereus* in the forest at MLBS.

H_2: That, in the laboratory there, adults of the latter species act aggressively toward those juveniles.

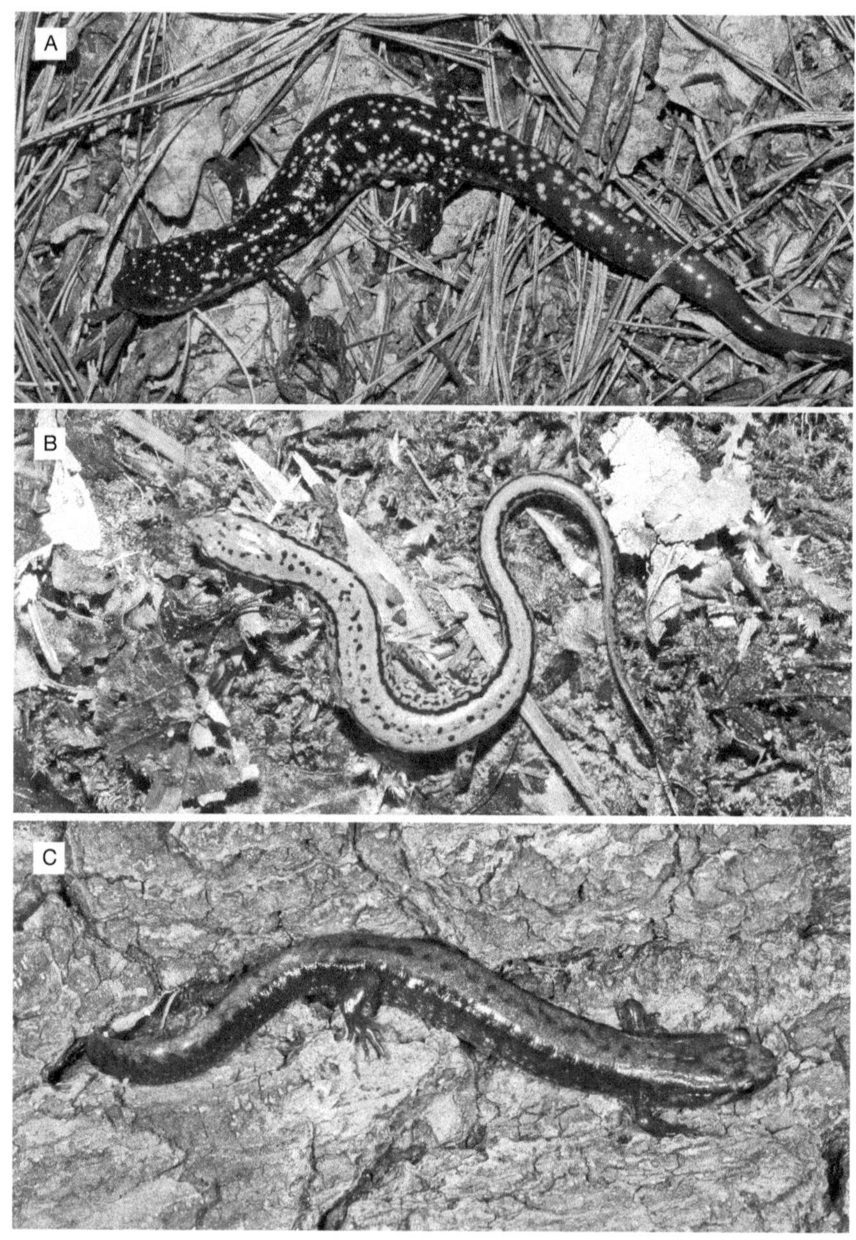

Figure 6.1 (A) Adult *Plethodon glutinosus.* (B) Adult *Eurycea cirrigera.* (C) Adult *Desmognathus fuscus.*

For 10 days, during June–July 1991 (noncourtship season) in a plot of forest, Lancaster and Jaeger (1995) turned 523 rocks and logs of sufficient sizes to be the foci of territories for adult *P. cinereus*. They found 336 of these housing only *P. cinereus*, 13 covering only juvenile *P. glutinosus*, zero covering both species, and 174 housing neither. This distribution was statistically compared to a random dispersion, and it was found that the two species cohabit significantly less often than expected by random chance ($p < 0.001$). However, conspecific juveniles shared 17% of adult *P. cinereus* territories while none of the cover objects were shared by the 20 adults of *P. glutinosus*. These data supported the a priori hypothesis that few (if any) juveniles of *P. glutinosus* are found in territories of *P. cinereus*. However, this nonrandom association could be caused by differential habitat preferences by the two species: that is, *P. cinereus* prefers to be under rocks and logs while juveniles of *P. glutinosus* do not.

Lancaster and Jaeger (1995) conducted a laboratory experiment at MLBS to test for interspecific aggression, which also might repel juveniles of *P. glutinosus* from territories of adults of *P. cinereus*. They collected adults of *P. cinereus* and juveniles of *P. glutinosus* near MLBS and chose individuals that intra- and interspecifically were close in head widths and SVL. Head widths determine the size of prey that can be ingested, so sympatric, morphologically similar salamanders should compete for prey. Each salamander was placed in a separate chamber, with *D. virilis* as prey, for 5 days (which is sufficient time for *P. cinereus* to establish a territory; Nunes & Jaeger, 1989, in section 3.8.1). Then each adult *P. cinereus* was observed for 15 minutes in each of three randomized conditions: with C_1: a conspecific adult intruder ($N = 24$); C_2: a juvenile *P. glutinosus* intruder ($N = 24$); and C_3: an odorless surrogate control (damp paper cylinder) as intruder ($N = 24$). Residents spent nearly equal time (mean responses) in aggressive behaviors (Fig. 3.1) toward both conspecific adults and the congeneric juveniles, which was significantly greater than the time spent in aggressive behavior toward the control. Also, four residents bit the conspecific adults and four bit the juveniles, but neither type of intruder bit the residents.

From these two studies, Lancaster and Jaeger (1995) inferred that (1) juveniles of *P. glutinosus* seldom occur in territories of adult *P. cinereus*, and (2) adults of *P. cinereus* act as aggressively toward those juveniles as they do toward adult conspecifics. They also suggested that the "rules of engagement" for *P. cinereus* are to be more aggressive toward same-size conspecific adults than toward conspecific juveniles (see Jaeger, 1995a, in section 7.1) and to be more aggressive toward same-size congeneric juveniles (this study) than toward congeneric adults (see Jaeger et al., 1998, in section 6.3).

6.3 *PLETHODON CINEREUS* IN AN ASSEMBLAGE OF SALAMANDERS

The salamander assemblage at MLBS consists of 13 species of which 11 are in the family Plethodontidae. Jaeger et al. (1998) were particularly interested in just four of the many plethodontid species that share the ecotone between streams (two semiaquatic species) and upland forest (two fully terrestrial species). Dr. Henry Wilbur (then director of MLBS) and his students were intensively studying the ecological relationships among the many ecotonal species at MLBS, so Jaeger et al. chose to study behavioral interactions of adults of *P. cinereus* with adults of *P. glutinosus* (both terrestrial upland species), *Desmognathus fuscus*, and *Eurycea cirrigera* (both breed in streams but the adults otherwise forage for various distances into the upland forest; Fig. 6.1A–C). Their three a priori hypotheses were

H_1: Because adults of *P. glutinosus* are much larger than adults of *P. cinereus*, the former acts as a predator of the latter.

H_2: *D. fuscus* is also a predator because it too is larger and more robust than *P. cinereus.*

H_3: *E. cirrigera* and *P. cinereus* are competitors for prey because they are approximately the same adult sizes, assuming that they utilize the same types of prey.

The purpose of these behavioral studies was to generate predictions about the interactive distributions of these (and other) species in later ecological studies on the forest floor within the stream-upland ecotone (for such ecological studies, see Ransom & Jaeger, 2006, 2008, in sections 6.4 and 6.5).

After collecting the four species at MLBS, Jaeger et al. (1998) placed each animal in a separate chamber in a laboratory at MLBS until the six experiments began (Fig. 6.2). Each very large *P. glutinosus* was fed one earthworm per week while the other species were fed *D. melanogaster* and *D. virilis ad libitum.* This allowed sufficient time for *P. cinereus* (and perhaps the other species) to establish territories with a rich supply of prey. The six experiments, with no prey present, examined whether *P. cinereus* ($N = 19$) is attacked by *P. glutinosus* ($N = 19$) as a predator (E_1 and E_2); *P. cinereus* ($N = 28$) is attacked by *D. fuscus* as a predator (E_3 and E_4); and *P. cinereus* ($N = 26$) is antagonistic with *E. cirrigera* as a territorial competitor (E_5 and E_6) (unequal sample sizes reflected difficulties in collecting the species, except for the ubiquitous *P. cinereus*).

Each experiment had three conditions. Experiment 1's conditions were C_1: a resident *P. glutinosus* with an intruding *P. cinereus*; C_2: a resident *P. glutinosus*

Figure 6.2 Caitlin Gabor in front of the Research Building at MLBS. All behavioral experiments at MLBS were conducted in the cool, dimly lit basement of this building.

with an intruding damp surrogate (size of *P. cinereus*); and C_3: a resident surrogate (size of *P. glutinosus*) with an intruding *P. cinereus*. Experiment 2 reversed experiment 1 exactly but with C_1: *P. cinereus* as resident/*P. glutinosus* as intruder; C_2: resident *P. cinereus*/intruding surrogate; and C_3: resident surrogate/intruding *P. glutinosus*. Experiments 3 and 4 repeated the previous methodology (three conditions each) but with E_3: *P. cinereus* as resident but in C_3 a conspecific as an intruder, and E_4: *D. fuscus* as resident but in C_3 *D. fuscus* as an intruder. Experiments 5 and 6 were identically designed but with E_5: *P. cinereus* as resident but in C_3 a conspecific as intruder, and E_6: *E. cirrigera* as resident but in C_3 another *E. cirrigera* as intruder.

The results did not reject the null hypothesis for E_1 and E_2, with nonsignificant results (among the three conditions) for the ATR threat posture; escape behavior; and zero snaps at, bites, or attempted ingestions. Both the small *P. cinereus* and the very large *P. glutinosus* seemed to ignore each other as much as they ignored the control surrogates, except that *P. cinereus* chemoinvestigated (nose tapped [NT]) the congener significantly more than the surrogates. A few individuals of *P. cinereus* even chose to lie on the heads of the congeners! Jaeger et al. (1998) inferred that *P. glutinosus* is not a predator on *P. cinereus*, counter to the original hypothesis. Also, while *P. cinereus* is aggressive toward same-size juveniles of *P. glutinosus* and excludes them from territories (Lancaster & Jaeger, 1995, in section 6.2), it does not do so with the adults.

Experiments 3 and 4 found dramatically different results. In E_3, residents of *P. cinereus* were significantly less aggressive toward (ATR, $p < 0.0006$), more submissive toward (FLAT, $p = 0.0018$), and spent more time moving away from (fleeing, $p = 0.0004$) *D. fuscus* than conspecific intruders. Similar results were found for E_4. In both experiments, *D. fuscus* attacked *P. cinereus* (e.g., 14 of 26 tests in E_3) and sometimes caused autotomization of their tails, which *D. fuscus* ate; *P. cinereus* never attacked *D. fuscus*. Also, *D. fuscus* similarly attacked its own conspecifics (e.g., 11 of 28 tests in E_4) but with no intraspecific cannibalism of tails. Although *D. fuscus* did frequently chase fleeing *P. cinereus*, as if attempting predation, for ethical reasons the observers (Jaeger and Gabor) did not allow predation to occur. Jaeger et al. (1998) inferred that *D. fuscus* does act aggressively toward *P. cinereus* and is a predator on it while the latter species flees from *D. fuscus* (which probably also acts as a predator toward its conspecifics). In another experiment in our research group at MLBS, Del Balso (1995) found that of 75 adults of *P. cinereus*, *D. fuscus* completely ingested 14.7%, caused tail autotomy in 17.3%, and wounded 12.0%, while 56.0% were unharmed. Indeed, *D. fuscus* is a dangerous predator on salamanders in the stream-upland ecotone.

Experiments 5 and 6 provided a striking behavioral departure from the results in E_1 through E_4. In E_5, intruding *E. cirrigera* showed significantly more move toward (MT) residents of *P. cinereus* than toward the control surrogate, but intruding *P. cinereus* did not do so in E_6. However, in E_5, residential *P. cinereus* violently bit some intruding *E. cirrigera*, which did not bite the residents. In E_6, intruding *P. cinereus* also bit a residential *E. cirrigera*, but the latter did not return this bite. Jaeger et al. (1998) inferred that their data only partially supported the hypothesis that these two species are territorial competitors. In support of the hypothesis, a residential *P. cinereus* in E_5 threatened (ATR) its intruding "rival" in 62% of encounters, pursued it in 2%, and sometimes violently bit it (but with no indications of attempted predation). However, *E. cirrigera* provided little evidence for inter- or intraspecific aggression or territoriality in either E_5 or E_6. This contrasts with the benign responses in the encounters between *P. cinereus*/*P. glutinosus* (E_1 and E_2) and predatory behavior by *D. fuscus* on *P. cinereus* (E_3 and E_4).

For later ecological experiments in the stream-upland ecotone at MLBS (see sections 6.4 and 6.5), Jaeger et al. (1998) made several predictions: (1) *P. cinereus* and same-size juveniles of *P. glutinosus* will be nonrandomly distributed in the ecotone, because the former apparently displaces these juveniles from territorial spaces (section 6.2); (2) adults of these two species will not affect each other's distributions, based on data in E_1 and E_2; (3) *D. fuscus* will be nonrandomly distributed intraspecifically due to cannibalism (E_3 and E_4);

(4) *D. fuscus* will cross the ecotone and displace *P. cinereus* by predation (E_3 and E_4); (5) where *P. cinereus* is removed by predatory *D. fuscus, E. cirrigera* (E_5 and E_6) will enter the ecotone from its stream breeding sites (unless it too is a prey for *D. fuscus*, for which there were no data) and replace otherwise territorial *P. cinereus;* and (6) overall, the distance that *D. fuscus* can penetrate into the upland forest will depend on two major factors: how far it can move away from its stream breeding habitat between courtship seasons and the extent of mortality it causes by eating *P. cinereus* and each other.

In sum, interspecific competition, predation, and possibly abiotic factors (e.g., rainfall) will interact extensively in determining the structure of the salamander community in the southern Appalachian Mountains. Next we turn to subsequent ecological experiments that tested some of those predictions.

6.4 ECOLOGICAL TESTS OF BEHAVIORAL PREDICTIONS: ENCLOSED PLOTS ON THE FOREST FLOOR

Ransom and Jaeger (2006) used forest mesocosms and laboratory behavioral experiments to test predictions derived by Jaeger et al. (1998) and to formulate new predictions concerning species-pair combinations in the stream–upland forest ecotone at MLBS. They installed 20 litter/soil-filled mesocosms (wood and screen enclosures, 1.22 m × 1.22 m × 10 cm tall) in shallow holes on the forest–stream interface: 10 as control and 10 as experimental units. Each of two experiments (summer 2005) was conducted twice (total of 40 mesocosms) to increase sample sizes: one set with *D. fuscus* and *P. cinereus* and the other set with *E. cirrigera* and *P. cinereus*. In each experimental mesocosm, the authors added one male *P. cinereus* and one male *D. fuscus* in the center, or similarly one male *P. cinereus* and one male *E. cirrigera*. For the controls, they placed two males of *P. cinereus* or two males of *E. cirrigera*. Six days later, they searched the mesocosms to determine mortality (predation) and the spatial distributions of the experimental (interspecific) salamander pairs versus the matched control (intraspecific) salamander pairs. Briefly, these results did not significantly support the prediction that *D. fuscus* is a successful predator on *P. cinereus* and also did not significantly support the prediction that *P. cinereus* avoids (flees from or remains spatially distant from) *D. fuscus*. Also, the data did not support the prediction that *P. cinereus* prevents *E. cirrigera* from moving from the streamside into the forest (also see Ransom & Jaeger, 2008, in section 6.5, for a similar result).

The results from Jaeger et al. (1998) versus Ransom and Jaeger (2006) illustrate that while simple laboratory behavioral experiments are useful in making predictions on interspecific associations, those predictions may or may not be supported by experiments within the salamander community on the complex forest floor. The natural habitat seems to provide the complexity (leaf litter, rocks, logs, fossorial retreats) needed for salamanders to avoid predation or to avoid competition in some situations. Therefore, one should beware of interpretations of behavioral results from the laboratory without more realistic studies in the heterogeneous natural habitat.

The salamander assemblage at MLBS contains five stream-breeding plethodontid species, besides *D. fuscus*, that may be predators as adults on *P. cinereus*, *E. cirrigera*, and/or each other. The very large adults of *Gyrinophilus porphyriticus* and *Pseudotriton ruber* were rarely found (about one adult each per year) on the forest floor while *D. monticola* and *D. quadramaculatus* were difficult to collect in and next to streams. The remaining species is *D. ochrophaeus*, which Ransom and Jaeger (2006) collected in sufficient numbers, so they designed two laboratory experiments at MLBS to test for predation or aggression by adults of this species and by *D. fuscus* on adults of *E. cirrigera* during 2005. These experiments were intended to generate predictions for later experiments in the stream–upland forest ecotone in section 6.5.

The experimental design was the same as Jaeger et al.'s (1998), except that six (instead of three) randomized conditions were employed. In E_1, they were C_1: *E. cirrigera* resident/*D. ochrophaeus* intruder; C_2: *E. cirrigera* resident/surrogate intruder; C_3: surrogate resident/*D. ochrophaeus* intruder; C_4: *D. ochrophaeus* resident/ *E. cirrigera* intruder; C_5: *D. ochrophaeus* resident/surrogate intruder; C_6: surrogate resident/*E. cirrigera* intruder. In E_2, *D. fuscus* was used in place of *D. ochrophaeus.*

In E_1, the results did not provide strong evidence that *D. ochrophaeus* was exceptionally aggressive toward *E. cirrigera*. Also, there were no significant differences in the behavior by the latter species toward the former species versus the surrogate controls, with the exception that *E. cirrigera* spent less time in the threat posture (ATR) as a resident than as an intruder in the presence of *D. ochrophaeus.* In E_2, *E. cirrigera* spent significantly more time in escape behavior (EDGE) when encountering *D. fuscus* versus the surrogates and significantly more time in escape behavior as an intruder than as a resident. This implied that *E. cirrigera* may avoid encounters with *D. fuscus* by fleeing. However, both *D. ochrophaeus* and *D. fuscus* infrequently bit *E. cirrigera*, which led to a surprising observation: the biters would next open their mouths widely and repeatedly as if trying to cleanse their mouths, a behavior never seen after *Desmognathus* bit a *P. cinereus* or after eating flies. This post-biting behavior suggested to Ransom and Jaeger (2006) that *E. cirrigera* may be unpalatable to *Desmognathus.*

The data provide evidence that *E. cirrigera* avoided the larger *D. fuscus* but not the smaller *D. ochrophaeus*, as if perceiving the former, but not the latter, as a threat. Possibly *E. cirrigera* (via ATR posture) and the not much larger *D. ochrophaeus* are competitors for invertebrate prey on the forest floor, but this scenario was beyond the scope of these behavioral tests. Overall, the predictions were that, in the ecotone, neither *D. fuscus* nor *D. ochrophaeus* would have a dramatic impact on the distribution of *E. cirrigera* on the forest floor.

6.5 MORE ECOLOGICAL TESTS OF BEHAVIORAL PREDICTIONS: UNENCLOSED PLOTS ON THE FOREST FLOOR

Ransom and Jaeger (2008) continued their ecological experiments in the stream–upland forest ecotone at MLBS by establishing 40 plots along Hunters Branch and Hogskin Creek in the New River drainage system (1160 m elevation). Their marked, but unenclosed, plots were 5 m wide perpendicular to and crossing the streams, extending 5 m from one edge and 15 m from the opposite edge. The terrestrial *P. cinereus* seldom crosses streams, but the semiaquatic *Desmognathus* and *Eurycea* do. The researchers removed *P. cinereus* from 10 plots, all four species of *Desmognathus* from 10 plots, both *P. cinereus* and *Desmognathus* from 10 plots, and no salamanders from 10 control (but equally disturbed) plots. All plots were separated along the streams by ≥20 m (allowing salamanders to move out of or into plots), and the four treatments were randomly distributed among the 40 plots. The removed salamanders were kept healthy in laboratory chambers and later returned to their plots at the end of the experiment, which lasted from May to August 2005.

The primary purpose of this study was to monitor the distributions of *E. cirrigera* as influenced by the removal of just *P. cinereus*, just *Desmognathus*, both, or neither. (Recall that in the laboratory, Jaeger et al., 1998, had behaviorally studied interactions of *P. cinereus* vs. *E. cirrigera* while Ransom and Jaeger, 2006, had done the same with *D. fuscus/D. ochrophaeus* vs. *E. cirrigera*.) Ransom and Jaeger (2008) searched all plots every 7 to 10 days for 10 total searches. During the first two and last two searches, they recorded for each salamander its species, mass, SVL, and distance from the streams' edges. During all 10 searches, appropriate incoming salamanders were removed depending on treatment. The authors found that *P. cinereus* and *Desmognathus* did not significantly affect one another either in terms of numbers of individuals found on removal versus control plots or in their distances from streams.

Because Jaeger et al. (1998) and Del Balso (1995) had found in laboratory experiments that at least *D. fuscus* is a predator on *P. cinereus*, these results from Ransom and Jaeger (2008) again suggest that the complexity of the forest floor habitat reduces the predatory success of the larger species of *Desmognathus* on

P. cinereus. Also, the removal of *Desmognathus* from plots, compared to the controls, had no significant effects on either the numbers or distributions (from streams) of *E. cirrigera*. This result had been predicted by Ransom and Jaeger (2006) in section 6.4 who had suggested from laboratory experiments that *E. cirrigera* is unpalatable to both *D. fuscus* and *D. ochrophaeus* (a suggestion supported by Del Balso, 1995). The distances of *E. cirrigera* from the streams, however, were significantly affected by the removal of *P. cinereus* compared to the control plots. Opposing the a priori predictions by Jaeger et al. that *E. cirrigera* would move farther upland from streams in the absence (removal) of aggressive *P. cinereus,* Ransom and Jaeger (2008) found that *E. cirrigera* moved significantly closer to streams with the removal of *P. cinereus*! They also found *E. cirrigera* significantly closer to streams in plots (vs. controls) where both *P. cinereus* and *Desmognathus* had been removed.

Concurrently, in plots where *P. cinereus* had been removed, the later invading (and then removed) individuals of *P. cinereus* were significantly smaller both in SVL and mass compared to the original inhabitants. This would be expected if smaller, nonterritorial individuals (the "floating population") invaded territories of the removed larger adults, as had been found by Mathis (1990a, 1991b) in section 3.7. How the smaller invading *P. cinereus* elicited a negative (closer to streams) response by *E. cirrigera* (if there were even such a cause and effect) led to wild speculation by Ransom and Jaeger (2008), which we wisely do not repeat here.

Ransom and Jaeger (2008) concluded that "interactions among salamanders across the stream–forest ecotone influences intergeneric responses in a complex fashion" (30). They also warned that "disturbances [e.g., in the laboratory, mesocosms, and plots] thought to affect only a single species actually can have unforeseen effects on other members of the greater community" (30). This warning, of course, was borrowed from Heisenberg's (1958) famous "uncertainty principle" in physics, that any attempt to discover both the position and the momentum of a particle is impossible due to the disturbances caused by the human observations or experiments (summarized by Isaacson, 2007). This is a warning to be taken seriously by us and other biologists when conducting either behavioral laboratory or ecological field experiments.

6.6 CHARACTER DISPLACEMENT: *P. CINEREUS* VERSUS *P. HOFFMANI*

Character displacement occurs when two or more species are similar in some set of traits when in allopatry but diverge in those traits when in sympatry; often the traits are morphological, such as SVL and head width in salamanders.

Theoretically, this divergence is a consequence of natural selection due to interspecific competition between sympatric, closely related species, such as between same-size species of *Plethodon*. Such displacement would reduce interspecific exploitative competition for a commonly utilized resource such as prey, leading to niche partitioning in sympatry. Such selection would not occur in allopatric populations of those species. An alternative hypothesis is that when the two similar-size species first encounter each other during range expansions, they compete by aggressive interference competition for that resource; this leads to later character displacement in areas of sympatry via alpha selection. For example, one species may grow to a larger size, thus enhancing its aggressive success (RHP), while the other species either does not change morphologically or becomes smaller in size. Both theories assume that character displacement is an evolutionary response to interspecific competition for some scarce resource. See Jaeger, Prosen, et al. (2002) for an elaboration of these theories.

Dean Adams (Iowa State University) has extensively studied character displacement among sympatric/allopatric species pairs of *Plethodon*. Jaeger, Prosen, et al. (2002) were especially interested in character displacement between *P. hoffmani* and *P. cinereus*. The former species occupies a large range within the more extensive distribution of the latter species, but, except for a small area of sympatry in northcentral Pennsylvania, *P. cinereus* is absent from the range of *P. hoffmani*. Previous morphological comparisons by Adams had found that SVL and six measures of head morphology were not different in sampled allopatric populations but were in the sympatric population: *P. hoffmani* was significantly larger in SVL (about 1.3 times) and significantly larger in head characteristics (1.2 to 1.3 times) than *P. cinereus* (Jaeger, Prosen, et al., 2002). Note that the larger head width of *P. hoffmani* may lead to food niche partitioning or superior aggressive biting ability in sympatry.

Jaeger, Prosen, et al. (2002) performed two laboratory experiments at ULL to test for allomonal (interspecies chemical) communication and aggressive behavior between the two species. They made four a priori predictions. Sympatric *P. cinereus* will (1) respond significantly more intensely to allomones of and (2) be significantly more aggressive toward *P. hoffmani* than will allopatric *P. cinereus*; or sympatric *P. hoffmani* (3) will respond significantly more intensely to allomones of and (4) be significantly more aggressive toward *P. cinereus* than will allopatric *P. hoffmani*. They posited that these predicted responses would be compatible with the hypothesis that alpha selection (via aggression) had induced the divergent morphology of one or both of the species in sympatry.

For the allomonal experiment, each allopatric and sympatric *P. hoffmani* and *P. cinereus* was placed in a separate Nunc bioassay chamber containing a damp paper towel and was fed *D. virilis* for 5 days so that it could "mark" the towel with bodily odors and fecal pellets. These were the "allomonal donors." Then each towel was rolled into a cylinder and placed in an identical chamber housing recipient allopatric and sympatric *P. hoffmani* and *P. cinereus*, which had been fed and established residency there for 5 days. This created eight test conditions for the recipients, and each condition was replicated only 10 times due to the paucity of sympatric *P. hoffmani* that had been collected. After the usual habituation time of 15 minutes for each recipient, Jaeger, Prosen, et al. (2002) recorded, for 15 minutes, the recipient's responses to the soiled towels: number of NT (chemoinvestigation) and total times spent in ATR, FLAT, and touching the cylindrical towel. Statistical comparisons were made, when response sample sizes were suitable, between how allopatric versus sympatric *P. cinereus* responded to marked towels of allopatric and sympatric *P. hoffmani* and vice versa.

Of the 24 suitable statistical comparisons, only 3 were significant. These results were hardly more than the probability of finding a significant difference by random chance alone. This might suggest either that (1) the two species do not produce allomones as interspecific signals between them, or (2) if they do, then they do not respond dramatically toward such signals. This contrasts with evidence that *P. cinereus* and *P. shenandoah* do respond mutually, usually negatively, to allomones of each other (see section 2.2). In the aggression experiment, the design was similar to that used in the allomonal experiment except that live residents/intruders of allopatric/sympatric *P. hoffmani* and allopatric/sympatric *P. cinereus* were tested pair-wise along with the surrogate controls (unmarked, cylindrical paper towels). Each of these 12 conditions was replicated 10 times and behavioral variables were recorded for 15 minutes as NT, ATR, FLAT, EDGE, touch, and bite. These behaviors were recorded separately for both residents and intruders.

The 42 meaningful statistical results were partitioned into three groups: comparisons that included (1) each species versus the surrogate control ($N = 21$ statistics), (2) only *P. cinereus* as residents versus intruding *P. hoffmani* ($N = 10$), and (3) only *P. hoffmani* as residents versus intruding *P. cinereus* ($N = 11$). We do not attempt to untangle this plethora of statistical results but instead focus on the main conclusions drawn by Jaeger, Prosen, et al. (2002). They found no support for the prediction that the allomones of *P. hoffmani* influence the behavior of *P. cinereus* and only weak support that the allomones of *P. cinereus* influence the behavior of *P. hoffmani*. In the latter case, sympatric *P. hoffmani* spent significantly more time in the FLAT posture than did allopatric conspecifics in response to allomones of allopatric *P. cinereus*.

The aggression experiment, however, showed that *P. cinereus* was much more violent than *P. hoffmani*. Both allopatric and sympatric *P. cinereus* were significantly more aggressive and less submissive than either allopatric or sympatric *P. hoffmani*. Also, *P. cinereus* as allopatric residents was significantly more threatening (ATR) than conspecific sympatric intruders and significantly more likely to touch the congeneric opponent than allopatric intruders. As intruders, allopatric and sympatric *P. cinereus* were significantly more threatening and less submissive than both allopatric and sympatric residents of *P. hoffmani* and significantly more likely to touch the opponent than allopatric residents of *P. hoffmani*. Finally, significantly more individuals of *P. cinereus* bit the congener than did *P. hoffmani* ($p = 0.0101$), which never bit. In summary, Jaeger, Prosen, et al. (2002) suggested that "*P. cinereus* is a very active and aggressive species whereas *P. hoffmani* is a very lethargic species" (398).

Jaeger, Prosen, et al. (2002) were puzzled by these results relative to the evolution of character displacement, because even though *P. cinereus* is the more aggressive of the two species (predicted under alpha selection), it is also the smaller of the two (not predicted under alpha selection). They asked (1) is the highly aggressive *P. cinereus* replacing the more docile *P. hoffmani* by moving into its large range or (2) might *P. hoffmani* balance this interference competition from *P. cinereus* by being better at exploitative competition for scarce prey? If the former is correct, then over a long period of time *P. cinereus* should extend its distribution at a cost to *P. hoffmani*; if the latter is correct, then the two species may have established a static boundary of contact between them caused by a trade-off between interference and exploitative competition. Clearly many further investigations await future researchers.

6.7 COMPETITION BETWEEN *P. CINEREUS* AND *P. HUBRICHTI*

While *P. cinereus* has a vast distribution in forests of northeastern North America, *P. hubrichti* (completely surrounded by *P. cinereus)* is endemic to a forest that is only about 8 × 15 km in area at Peaks of Otter in the Blue Ridge Mountains of western Virginia. Both closely related species inhabit apparently similar mature hardwood forests; forage under rocks, logs, and leaf litter; are of similar adult sizes (*P. hubrichti* 39.6–56.9 mm SVL; *P. cinereus* 35.5–52.3 mm SVL at Wicknick's, 1995, site); and are parapatric over an area of 50 to 150 m. (Note that this area of parapatry is much larger than that for *P. cinereus* and *P. shenandoah* discussed in section 2.1.) The two species are not known to hybridize and appear to be ecological equivalents. Given this, Wicknick tested the hypotheses that *P. cinereus* and *P. hubrichti* engage in interspecific competition and territoriality. One other hypothesis (Highton, 1971), not directly tested by Wicknick, is that *P. cinereus* is so aggressive that it is slowly invading the habitat of *P. hubrichti* and replacing it there, causing another relict species of *Plethodon* to become extinction-prone. This scenario is similar to the dominance of *P. cinereus* over *P. shenandoah* (Griffis & Jaeger, 1998; Jaeger, 1970).

6.7.1 Habitat niche partitioning?

Wicknick (1995) first tested the alternative hypothesis that the distributions of these two species are merely due to preferences for different habitats. If so, then ongoing interspecific competition need not occur and the area of parapatry would be just the ecotone between the different allopatric habitats. She conducted soil and floristic surveys in the area of parapatry and in both allopatric areas. For the soil survey, she measured soil moisture, pH, and depth at 20 randomly chosen sites in each area. For the floristic survey, she recorded data for 80 trees in each area and calculated relative density, frequency, and coverage at basal area of the species; these values combined to produce the "proportional similarities" among the three areas.

Soil moisture (a critical factor for *Plethodon*) did not vary among the three areas. Soil pH was significantly more acidic at the allopatric area of *P. cinereus* than at the sympatric area; soil depth was significantly deeper at the sympatric area than at either of the allopatric areas. The floristic survey found a total of 22 species of trees with much species' overlap among the three areas: 15 species for the allopatric *P. hubrichti*, 13 in the sympatric area, and 11 for the allopatric *P. cinereus*. Proportional similarity was 36.25% between areas of sympatry and allopatric *P. hubrichti*. The two allopatric areas were similar by 28.75%, while the allopatric/sympatric areas of *P. cinereus* were similar by 26.25%. For all areas combined, similarity was 22.50%. Wicknick (1995) inferred that no obvious differences in soil characteristics occurred between the two allopatric areas. Also, plant communities in all three areas consist of hardwood species of trees with Northern Red Oak as the most important species. These species of hardwoods are known to dominate forests in which various species of *Plethodon* thrive. Therefore, Wicknick concluded that no apparent, gross habit distinctions occurred that would explain the allopatric distributions of *P. cinereus* and *P. hubrichti*.

6.7.2 Microhabitat niche partitioning?

If differences in gross features of the two species' areas do not lead to allopatry, perhaps differences in the species' use of microhabitats do. Wicknick (1995) divided this research into four studies.

First, in the area of sympatry, Wicknick (1995) spent two rainy/foggy nights locating salamanders, in a transect, that were foraging on top of the leaf litter. She did the same during three day searches by turning rocks and logs. She counted the number of salamanders of each species and uniquely marked each individual to avoid double-counting in the sympatric area. The results showed no significant difference in the ratio of numbers of the two species during the two surveys. She concluded that neither species dominated cover objects in sympatry and that both species were found in similar microhabitats.

Second, for 10 days, during September 1993 (courtship season), Wicknick (1995) turned 327 cover objects during daylight and recorded 352 salamanders under them, partitioned among the two allopatric and the sympatric areas. A few male–female pairs were found. She recorded sex and species for each adult and just species for the subadults (neonates and juveniles). Subadults were not included in the analyses, but their presence indicated that the species bred in both allopatric and sympatric areas. Positions of the cover objects were mapped in each area and salamanders were released at the sites of capture.

She found 68 adults under 78 cover objects in the allopatric area of *P. hubrichti*, 71 adults per 79 cover objects in the allopatric area of *P. cinereus*, and in sympatry 93 adults of *P. hubrichti* per 103 cover objects and 60 adults of *P. cinereus* per 73 cover objects. Males and females of *P. hubrichti* were each distributed perfectly uniformly with respect to sex in both allopatry and sympatry as was *P. cinereus* in sympatry and nearly so in allopatry. Wicknick (1995) inferred that the uniform distributions within sexes indicated that both sexes of both species are territorial. Negative intersexual spatial associations indicated that territoriality occurs between sexes too. Negative interspecific spatial associations indicated that either (1) the two species are interspecifically territorial or (2) a microhabitat difference occurs between the two species. Option 1 was supported by the unsegregated interspecific spatial distribution. Therefore, Wicknick again found no evidence for preferences for different microhabitats between the species but did find tantalizing spatial evidence for intra- and interspecific territoriality.

In her third study, Wicknick (1995) predicted that if option 2 were correct, then adults of *P. cinereus* and *P. hubrichti* may prefer to be under cover objects that differ in microhabitats by fine details. During early June (end of courtship season) after a rainfall, Wicknick minutely searched 30, 1 m^2 quadrats in randomly placed transects in both allopatric areas and in sympatry. She recorded species, SVL, and precise capture location of each salamander in each quadrat and for each measured the percentage of area covered by leaf litter and by each cover object. Salamanders were then released into their respective quadrats. She found no significant differences between species in usage of quadrats' microhabitats among the allopatric and sympatric areas. She thus inferred that the distributions of the two species in the leaf litter and under cover objects were not different from random. This did not support option 2 that the two species differ by fine-grain microhabitat usages. (For methods used in sampling populations in natural habitats, as Wicknick did, see Jaeger and Inger, 1994, for quadrat sampling; Jaeger, 1994b, for transect sampling; and Jaeger, 1994a, for patch [area] sampling.)

Wicknick's (1995) fourth study of species' distributions was to test the null hypotheses that neither *P. cinereus* nor *P. hubrichti* differs in body size between allopatric and sympatric areas. She collected 533 adults from under rocks in all three areas and recorded for each its species, sex, and SVL. Within allopatric areas, there were no significant differences in SVL between the sexes for either *P. hubrichti* or *P. cinereus*. In the sympatric area, adults of *P. hubrichti* were significantly larger than their conspecifics in the allopatric area, and the same was found for *P. cinereus*. Comparing the two species, adults of *P. hubrichti* were significantly larger than adults of *P. cinereus* in both allopatry and sympatry.

Wicknick (1995) suggested that if interspecific competition occurs, *P. hubrichti* should have a competitive advantage over *P. cinereus* in sympatry because of its larger size (i.e., greater RHP). (She tested this possibility later.) Curiously, both species were larger in sympatry than either was in allopatry. This suggests to us another case of character displacement such as between *P. cinereus* and *P. hoffmani* (Jaeger, Prosen, et al., 2002, in section 6.6). However, at Peaks of Otter this seeming "character displacement" is unclear for two alternative reasons. First, natural selection may have led to genetic factors underlying the larger sizes in sympatry either by alpha selection for enhanced aggression or for food niche partitioning of prey by taxa and sizes, and second, Wicknick (1995) had found that the sympatric area had deeper soil than either allopatric area. Deeper soil might then lead to prey of either higher quality for salamanders or of greater quantity (density) of prey. If the second alternative were true, then the seeming "character displacement" could merely be caused by both species growing larger in sympatry due to enhanced diets there. Unfortunately, Wicknick did not census the prey among allopatric and sympatric areas, a task left for future evolutionary ecologists.

6.7.3 Intra- and interspecific competition?

The hypothesis of competition was supported by the censuses of salamanders in the forest. The two species appeared to utilize identical microhabitats, and the intrasexual, intersexual, and interspecific distributions suggested both intra- and interspecific territoriality. Therefore, Wicknick (1995) conducted two experiments to test for competition.

First, she tested competition for artificial cover objects ("territories") in the laboratory at ULL. She size-matched paired individuals (which reduced asymmetric RHP) in both intra- and interspecific trials for their abilities to inhabit under a larger stone tile (superior site) or a smaller tile (inferior site) in the chambers. Before the experiment, each salamander spent 16 days alone in a chamber to become familiar with the tiles, and 90% chose the larger tile.

The experiment began by placing two salamanders in a chamber simultaneously, which eliminated residency (RHP) status. One was a randomly chosen as the "focal" salamander, for which data were recorded, and the other was its "opponent," for which data were not recorded. The focal and opponent were then switched in random sequences. Both intra- and interspecific tests were included, during which Wicknick (1995) recorded only which salamander obtained the superior tile and which did not. She found that both salamanders shared the superior tile in 60% to 80% of the tests in both intra- and interspecific conditions. Consequently, neither *P. hubrichti* ($p > 0.099$) nor *P. cinereus* ($p > 0.99$) had a competitive advantage in dominating the superior tiles. Four alternative inferences can be drawn from these results. (1) The two species are equally competitive for cover objects, (2) they do not compete for cover objects, (3) the reductions of differences in RHP by size and residency led to no superiority by one contestant over the other, or (4) the experimental design was flawed. Therefore, more experiments were needed to resolve these divergent inferences.

Wicknick (1995) then devised a competitive release experiment in the forest. She located under rocks, during the autumn 1993 courtship season, *P. hubrichti* ($N = 64$) in its allopatric area, *P. cinereus* ($N = 67$) in its allopatric area, and both *P. hubrichti* ($N = 72$) and *P. cinereus* ($N = 50$) in sympatry. Each rock was flagged; each salamander was recorded for species, sex, and SVL; and each was uniquely marked. The rocks were equally assigned as either experimental or control condition in each of the three areas and for each species. Salamanders under control rocks were released where found, and those under experimental rocks were displaced >100 m to inhibit homing behavior. Wicknick then repeatedly checked the rocks later in autumn 1993 and again during autumn 1994. For salamanders that subsequently invaded control and experimental rocks, she recorded species, time to invasion, and SVL.

In the allopatric areas, *P. hubrichti* invaded experimental rocks significantly faster than control rocks, but *P. cinereus* invaded controls significantly faster than experimental rocks. In neither allopatric area were their significant differences in sizes of conspecifics invading control versus experimental rocks. In sympatry, no significant differences were found for invasion times either between species or between each species' sympatric and allopatric populations. In the sympatric area, two significant differences were found for invasion frequencies. Areas occupied by *P. cinereus* (the controls) were invaded significantly more often by conspecifics than by *P. hubrichti*. Also, *P. cinereus* invaded rocks (control and experimental combined) significantly more often than did *P. hubrichti*.

Wicknick (1995) inferred that, in allopatry, *P. hubrichti* conformed to the hypothesis of intraspecific competition for cover objects but *P. cinereus* did not. She suggested that the results for *P. cinereus* were due to courtship seasons when males and/or females would be searching for mates; 58% of the invasions under control rocks were intersexual. In sympatry, *P. cinereus* was a better invader than *P. hubrichti*. This suggests that *P. cinereus* may have a slight competitive advantage over *P. hubrichti* in their quests for cover objects.

6.7.4 Intra- and interspecific territoriality?

Here Wicknick (1995) tested for territoriality (both intra- and interspecific) by adopting Jaeger and Gergits' (1979, in section 3.1) four criteria for *P. cinereus*: site fidelity (both site attachment and homing to the site), aggressive defense of the site, expulsion of intruders, and advertisement of the site. We review these in that sequence of her tests.

First, the test for site attachment was conducted in the forest at Peaks of Otter during autumns of 1993 and 1994. Wicknick (1995) located allopatric *P. hubrichti* (N = 64), allopatric *P. cinereus* (N = 67), and in sympatry both *P. hubrichti* (N = 72) and *P. cinereus* (N = 50). Each cover object was flagged and each adult salamander uniquely marked and placed back under its cover object. All cover objects were repeatedly turned during both autumns. Significantly more salamanders were recaptured under their original cover objects than expected by random chance for both species and in each of the allopatric and sympatric areas. There were no significant differences in recaptures at home sites for either species when comparing allopatric versus sympatric populations. For the second year (1994), she again found significantly more recaptured individuals than expected by random chance. She inferred that both species are not only faithful to their cover objects but that this faithfulness lasts for at least 13 months (i.e., the duration of her study). Therefore, both species are indeed philopatric.

Second, Wicknick (1995) conducted the test for homing ability only for allopatric *P. hubrichti*, because homing by *P. cinereus* had been well-documented (Kleeberger & Werner, 1982, in Michigan; later by Jaeger et al., 1993, at MLBS). She located 26 adults under cover objects during October 1994 and partitioned them into control (N = 13) and experimental (N = 13) conditions. All cover objects were flagged and salamanders uniquely marked; the controls were returned to under their cover objects whereas the experimental individuals were displaced 5 m away in random compass direction. Wicknick examined the cover objects 11, 15, 16, 17, and 19 days later. She compared the number of recaptures between control and experimental conditions and found no significant difference between recaptures under control and experimental cover objects. This indicated that individuals of *P. hubrichti* do home to their original sites as had been previously found for *P. cinereus*. Therefore, both species seem to conform to Jaeger and Gergits' (1979, in section 3.1) first criterion for territoriality in *Plethodon*: site fidelity (philopatry) by remaining "at home" and by homing back to a site when displaced.

Recall that Jaeger and Gergits' (1979, in section 3.1) considered aggression to be necessary for a salamander to expel invaders from its territory. Intraspecific aggression by *P. cinereus* has been well documented (see chapter 3), and aggressive contests between *P. cinereus* and *P. shenandoah* (Griffis & Jaeger, 1998; Wrobel et al., 1980) and between *P. cinereus* and *P. hoffmani* (Jaeger, Prosen, et al., 2002) were intense. Therefore, in her third study, Wicknick (1995) staged 180 contests, in the laboratory at ULL, between and within *P. cinereus* and *P. hubrichti*, using 30 residents and 30 intruders of each species. Each resident was tested, in random sequence, against a conspecific, a congeneric, and a surrogate control during the noncourtship summer of 1994. Residents were fed alone in their chambers for 107 days so as to adjust to the laboratory and the chambers and to mark their substrates. On a test day, an intruder (or surrogate) was introduced and both salamanders were allowed to habituate for 5 minutes after their handling. She then monitored the behaviors of each resident and intruder for 15 minutes and recorded five aggressive behaviors (ATR, look toward, move toward, nip-bite, and gripping bite), three submissive behaviors (look away, move away, FLAT), and seven "other" behaviors (front of trunk raised, NT to substrate, NT to opponent, walk on, walk under the opponent, and partially or fully burrow into the substrate). Her goal was to estimate the overall aggressiveness and submissiveness of each species from Peaks of Otter. Recall from chapter 3 that in another laboratory, Quinn and Graves (1999) had inferred from their experiments that *P. cinereus* is neither aggressive nor territorial in northern Michigan, and Wise and Jaeger (2016) in section 3.10 found geographic variation in territorial behavior within *P. cinereus*.

Luckily for us, Wicknick (1995) lumped the total times in aggressive and, separately, submissive behaviors for ease in statistical analyses. Residents of *P. hubrichti* were significantly more aggressive in both intra- and interspecific tests than in control tests, but there was no significant difference in aggression toward conspecifics versus *P. cinereus*. Residents of *P. cinereus* were also more aggressive in both intra- and interspecific tests than in the control tests, but they were significantly more aggressive toward conspecifics than toward *P. hubrichti*. Residents of *P. hubrichti* were significantly more submissive in the presence of conspecifics than with surrogates, but they were not significantly different in submission toward *P. cinereus* versus surrogates or versus conspecifics. The same was found for residential *P. cinereus*.

Between species, residential *P. hubrichti* was significantly more aggressive than was residential *P. cinereus* in intraspecific tests; they were also significantly more aggressive than *P. cinereus* in interspecific tests. Submissive behaviors of residential *P. hubrichti* and *P. cinereus* were not significantly different in intraspecific tests, but residential *P. cinereus* was significantly more submissive in interspecific tests than was *P. hubrichti*. Both species bit in both intra- and interspecific tests. For *P. hubrichti*, mean numbers of bites were 9.8/test intraspecifically and 2.0/test interspecifically. For *P. cinereus*, they were 6.0 intraspecifically and 0.03 interspecifically. There were no significant differences in biting by residents in intra- or interspecific tests for either species.

Wicknick (1995) was clever enough to record where residents bit intruders. Jaeger (1981, in section 3.5) had reported that *P. cinereus*, at SNP, bit conspecifics significantly more on the NLG than on the tail or on the trunk, so Wicknick's experiment was informative as a comparison. Due to low number of bites, she grouped the data for intra- and interspecific contests. *Plethodon hubrichti* bit intruders on the tail (50.9% of bites), trunk (32.7%), head (10.9%), cloacal region that contains the postcloacal gland (3.7%), and NLG (1.8%). Bites by *P. cinereus* were to the cloacal region (48.0%), tail (24.0%), trunk (20%), and head (8.0%), but no bites were directed toward the NLG. Therefore, *P. cinereus* at Peaks of Otter appears to bite the NLG of opponents far less than do conspecifics at SNP.

Wicknick (1995) inferred that both *P. hubrichti* and *P. cinereus* met Jaeger and Gergits' (1979, section 3.1) criterion for aggressive defense of territories, because both species were more aggressive toward conspecific and congeneric intruders than toward surrogate controls. The most severe form of aggression for *Plethodon* is biting, especially on the opponents' vulnerable NLG and tails, and both *P. hubrichti* and *P. cinereus* did bite both conspecific and congeneric intruders. However, *P. hubrichti* aimed most of its bites toward the fat-filled tails (which a bite can cause to autotomize), while *P. cinereus* aimed most of its bites toward the region of the postcloacal gland (which is involved in producing territorial and other pheromones). In contrast, the critical NLG (used for chemodetection) were rarely bitten by *P. hubrichti* and never by *P. cinereus* from Peaks of Otter. Wicknick also inferred that residents of both species displayed little submission toward intruders of either species but were more submissive toward them than toward surrogate controls. This too would be expected during contests if territorial residents and intruders engaged in a "game" of testing each other by posturing, thus reducing a contest's escalation to severe injury (e.g., biting), as proposed by Maynard Smith and Parker (1976).

In sum, Wicknick (1995) concluded that *P. hubrichti* is more aggressive than *P. cinereus* at Peaks of Otter. She supported this conclusion by noting that, in contests, a larger number of residential *P. hubrichti* bit, and bit more frequently, than did residential *P. cinereus*. If Wicknick is correct, then her data would not support the hypothesis that *P. cinereus* is so aggressive (as it is in SNP) that it should slowly displace *P. hubrichti* from its small area of forest.

Recall that Jaeger and Gergits (1979, in section 3.1) considered expulsion of intruders as another criterion for territorial behavior by *Plethodon*. Their logic was that even if a species is aggressive, this may not necessarily lead to territorial expulsion of intruders, because such competitors may endure aggression so as to gain entry into a food- or mate-rich territory or even to share it with the aggressive resident. A confounding question is: Which types of intruders might a resident regard as "competitors"? For example, a resident of either sex might endure "intrusions" by mates, potential mates, and socially monogamous partners (topics explored in chapter 7). Also the resident might favor "intrusions" by close kin such as offspring (see section 7.18). Finally, familiar, nonthreatening territorial neighbors may be endured (as explored in section 9.4). Therefore, experiments for "expulsion of intruders" confront several methodological and philosophical problems. One solution would be to test residents and intruders from distant localities, which would eliminate these three problems, but to do so would introduce the new problem of geographic variation in territorial behavior, as Wise and Jaeger (2016, in section 3.10) found within *P. cinereus*.

Wicknick's (1995) experimental design for exclusion wisely ignored these confounding problems, but her design was more sophisticated than the one used in her previous "competition for artificial territories" experiment, which had yielded nonsignificant results. She used *P. cinereus* (N = 30) and *P. hubrichti* (N = 30) as both residents and intruders with three randomized conditions: C_1: *P. cinereus* as resident with either a conspecific or *P. hubrichti* as intruder; C_2: *P. hubrichti* as resident with either a conspecific or *P. cinereus* as intruder; and C_3: either species as resident but no intruder present (control). A resident spent 7 days in its chamber alone, which contained a moist substrate, a "superior" cover object (larger raised tile, kept wetter, with *D. melanogaster* under it) and an "inferior" cover object (smaller raised tile, kept drier, with no prey under it). The intruder was then introduced, and the positions of both salamanders were recorded (overnight) 12 hours later.

The results were that residents of neither species were expelled from superior cover objects by intruders. Wicknick (1995) inferred, that regardless of species, residents were able to expel (or just keep out) intruders of both species. Intruders of *P. hubrichti* were expelled (or kept out) from resident-occupied areas (i.e., near the superior cover objects) both by conspecifics and by *P. cinereus*. Intruders of *P. cinereus*, however, were significantly less likely to be expelled (or kept out) from such areas by residents of either species. Wicknick inferred that *P. cinereus* is a better (if only slightly) invader/defender of cover objects than is *P. hubrichti*.

Finally, Jaeger and Gergits (1979, section 3.1) claimed that a territorial *Plethodon* should advertise its territory, presumably by pheromones, as shown for *P. cinereus* in chapter 5. This would warn potential intruders against entering the site. Conversely, Gosling (1990) claimed that territorial scent marks by antelopes do not deflect intruders but instead invite them to identify and test the combat abilities of the actual resident. Some evidence supports Gosling's theory for *P. cinereus* as well (discussed in chapter 5). In either case, a territorial *Plethodon* would be expected to scent-mark its defended site.

Wicknick (1995) tested the territorial repulsion argument with both *P. hubrichti* and *P. cinereus*. She tested both species in four experimental conditions in which a long tube contained a burrow at each end with a pellet in front of it. The pellets were either a control (moistened black soil shaped like a fecal pellet) or an experimental fecal pellet (deposited by a salamander alone in its home chamber). The conditions were C_1: own pellet versus control pellet, C_2: a conspecific's pellet versus control, C_3: a congeneric's pellet versus control, and C_4: two control pellets (the general control condition). Each salamander was randomly tested as an intruder in each of the four conditions. Wicknick observed each salamander for 30 minutes for total time spent in each burrow, and sample sizes ranged from $N = 19$ to 27 responders (entered one or both burrows). Statistical alpha was set a priori at 0.017 due to the multiple comparisons.

Wicknick (1995) detected no significant responses to control versus salamander's pellets among any statistical comparison among conditions, and p values ranged from 0.069 to 0.833. Wicknick posed two a posteriori hypotheses to account for her results. Neither *P. hubrichti* nor *P. cinereus* at Peaks of Otter (1) produce territorial pheromones/allomones, or (2) if they do, they just ignore these signal markers when entering a burrow. Her results are in contrast to those of Jaeger and Gergits (1979), who had found significant intra- and interspecific pheromonal/allomonal communication by *P. cinereus* and *P. shenandoah* from SNP, which is north of Peaks of Otter in the Blue Ridge Mountains of Virginia (see section 2.2). Also, fecal pellets have been used to identify territorial advertisement and repulsion of intruders by both males (Jaeger et al., 1986, in section 5.2) and females (Horne and Jaeger, 1988, in section 5.3) of *P. cinereus* from MLBS in the Appalachian Mountains of Virginia. More studies are needed to investigate possible geographic variations in production of, detection of, or repulsion of intruders by territorial pheromones and allomones.

6.7.5 Summary

Highton (1971) suggested that the sibling species of *P. hubrichti*, *P. nettingi*, and *P. shenandoah* are relicts of a formerly widespread species and are surviving now only in marginal habitats. The last two species are confined to mountaintops, but *P. hubrichti* is in a high-quality forest, though in a tiny area (Wicknick, 1995). All three relicts have parapatric interfaces with *P. cinereus*, and Highton (1971) argued that they are being displaced by *P. cinereus*.

Table 6.1 SUMMARY OF WICKNICK'S (1995) RESULTS FOR INTRA- AND INTERSPECIFIC TERRITORIALITY (COMPETITION) WITHIN AND BETWEEN PC AND PH AT PEAKS OF OTTER, VIRGINIA

Test (inference)	Intraspecific territoriality		Interspecific territoriality	
	PC	PH	PC	PH
Occupy same forest habitat (no niche partitioning)	—	—	Yes	
Occupy same forest microhabitats (no niche partitioning)	—	—	Yes	
Adults within sexes are spaced uniformly in forest (competition)	Yes	Yes	Yes	Yes
Adults between sexes are spaced uniformly in forest (competition)	Yes	Yes	Yes	Yes
Species prefer same types of cover objects in forest (no niche partitioning)	—	—	Yes	Yes
Character displacement in sympatry (competition?)	—	—	yes (smaller species)	yes (larger species)
Adults occupy different cover objects in laboratory (competition?)	No	No	No	No
Adults exhibit competitive release in forest (competition)	No	Yes	Yes (superior invader)	Yes (inferior invader)
Adults show site attachment in forest (competition)	Yes	Yes	Yes	Yes
Adults home to own sites in forest (competition)	Yes (from other studies)	Yes	—	—
Adults are aggressive in laboratory (competition)	Yes	Yes	Yes (inferior aggressor)	Yes (superior aggressor)
Adult residents can expel intruders from home sites in laboratory (competition)	Yes	Yes	Yes (but superior invader)	Yes (but inferior invader)
Adults advertise home sites with pheromones/allomones in laboratory (competition?)	No	No	No	No

NOTE. PC = *P. cinereus*; PH = *P. hubrichti*.

From her colossal study, Wicknick (1995) concluded that both *P. hubrichti* and *P. cinereus* at Peaks of Otter are intraspecifically territorial (see summary evidence in Table 6.1). She also concluded that the two species are interspecifically competitive, but, depending on how one interprets her data, they are either (1) competitive equals or (2) *P. cinereus* has a slight competitive advantage over *P. hubrichti* (Table 6.1). If case 1 were true, then the two species may maintain a static boundary (sympatry) between their distributions. If case 2 were correct, then *P. cinereus*, however, should be slowly displacing *P. hubrichti*, as Highton (1971) had suggested. Wicknick ventured no conclusion concerning interspecific territoriality, but Table 6.1 suggests to us that some complex form of interspecific territoriality occurs at Peaks of Otter. While *P. hubrichti* is larger in sympatry and is a superior aggressor (both enhanced RHP), *P. cinereus* is a superior invader of cover objects in both forest and laboratory experiments. The solution hinges on whether the superior RHP of *P. hubrichti* is sufficient to inhibit invasions by *P. cinereus*.

6.8 DIVERSITY OF BEHAVIORS BY *P. CINEREUS* TOWARD OTHER SPECIES

We have by now summarized the diversity of agonistic behaviors by *P. cinereus* toward seven other species of salamanders with which it interacts in northeastern forests. Here we provide our generalized understanding of those behavioral patterns with the caveat that, under identical experimental conditions, individuals of *P. cinereus* demonstrate a large variance in all behaviors observed. The possible reasons for this variance are explored in section 10.1.

Eurycea cirrigera—*P. cinereus* showed the most extreme forms of aggression toward this species, by escalating threat postures to ATR 3, 4, and 5 (Fig. 3.1) followed by vicious biting attacks and occasional attempts at predation. This confamilial species showed little aggression in return, but it may prey on the eggs of *P. cinereus*.

Plethodon hoffmani, *P. hubrichti*, and *P. shenandoah*—*P. cinereus* was also very threatening (ATR 3, 4, and 5) toward these same-size species but with far fewer biting attacks than toward *E. cirrigera*.

Plethodon glutinosus (juveniles)—*P. cinereus* exhibited low levels of aggression (ATR 1 and 2) toward these same-size juveniles and seldom launched a biting attack.

Plethodon glutinosus (adults)—This is neither a predator on nor a competitor for prey or territories with *P. cinereus* because of its very large size; *P. cinereus* acted benignly toward it, usually remaining in the resting posture (FTR).

Desmognathus fuscus—This is a large predator on *P. cinereus*, which responded by moving away from it.

Desmognathus ochrophaeus—This species is smaller than *D. fuscus* but still may be an occasional predator on *P. cinereus*, which responded by moving away.

Overall, *P. cinereus* was most aggressive toward presumed competitors for food and territories and toward a presumed oophagous confamilial, but predatory salamanders (confamilial species) induced flight in laboratory experiments. In chapter 7, we further explore the diversity of intraspecific social behaviors.

6.9 SELECTED RECENT RESEARCH BY OTHERS: INTERSPECIFIC TERRITORIALITY

A question not addressed in this chapter is the relationship between territorial red-backed salamanders and invertebrate competitors. In many cases, individuals of *P. cinereus* share spatial and dietary resources with predatory invertebrates. Although little is known about whether such invertebrate species are territorial, it is likely that they enter salamander territories where they may compete for prey and/or space. A number of researchers have adopted criteria from Jaeger and Gergits' (1979, in section 3.1) list of requirements for territoriality and asked if individuals of *P. cinereus* treat intruding invertebrates as territorial invaders. This approach has been applied to the carabid beetle *Platynus tenuicollis* (Gall et al., 2003), the centipede *Scolopocryptops sexspinosus* (Hickerson et al., 2004), and the amaurobiid spider genus *Wadotes* (Figura, 2007). In all cases, resident salamanders in laboratory encounters behaved aggressively (when compared to a damp paper cylinder control) by approaching, displaying ATR, and in some cases biting intruding invertebrates. To examine the interactions of *P. cinereus* and invertebrates in the forest, Hickerson et al. (2004) and Figura (2007) tested for nonrandom associations on the forest floor between salamanders and presumptive invertebrate competitors. Both researchers found significant negative associations under cover objects (supporting the laboratory findings). Presence of one species seemed to preclude occupancy by the other.

An experimental field approach to the study of salamander–invertebrate interactions was undertaken by Hickerson et al. (2012). During a 4-year experiment, they removed thousands of salamanders from replicated cover-object arrays on a forest floor in northeastern Ohio. In addition to examining the role of red-backed salamanders in the diverse temperate forest food web, they found a response of spiders to salamander removal. Compared to control plots, spiders significantly increased on salamander removal plots, a result the authors attributed to interference competition between salamanders and spiders, rather than direct predation by salamanders on spiders. Also, salamanders increased in plots where centipedes were removed, raising the possibility that centipedes exclude salamanders from cover and/or outcompete them for available prey.

Because so many species of nonnative invertebrates have been introduced within the geographic range of *P. cinereus*, researchers have focused on the conservation implications of salamander-invertebrate interactions. Anthony et al. (2007) compared the responses of juveniles of *P. cinereus* to odors of native and introduced centipede species. They found that salamanders responded to the odors of both species of centipedes by adopting submissive postures. Juvenile salamanders were negatively associated with centipedes under cover on the forest floor, and in laboratory trials centipedes were able to exclude salamanders from cover objects. Despite the small size of juvenile salamanders used in this study, no instances of predation by centipedes on salamanders were observed. This suggests, again, that the types of interactions that occur between salamanders and large invertebrates might be competitive rather than predatory. Earthworms represent another nonnative group that has the potential to interact with salamanders by physically interfering with foraging, egg brooding behavior, or even territorial defense. Large-surface active worms of the invasive Asian genus *Amynthas* are good candidates for this type of direct interference because they occupy the same cover that salamanders use as territorial foci. Ziemba et al. (2015) paired adults of *P. cinereus* with large worms in laboratory microcosms containing multiple cover objects. Salamanders in worm treatments significantly increased their movement from cover to cover, a behavior that would put them at risk of desiccation, if not predation, as exhibited in the forest.

Studies of interspecific interactions necessarily leave out important social questions related to how different sexes interact or how adults behave toward related and nonrelated juveniles. In chapter 7, we address these questions as we explore the diversity of intraspecific social behaviors in *P. cinereus*.

7

Intraspecific social behavior within *P. cinereus*

In chapter 3 we discussed one aspect of social behavior, namely intraspecific territoriality by *P. cinereus*. The observation of a larger intersexual versus intrasexual overlap of territories in this species (Mathis, 1991b, in section 3.7) led to more detailed studies of the spatial distribution of salamanders in their natural habitat. We describe here the many studies on the behavioral associations and interactions between adult conspecific salamanders, of the same or opposite sex, and the associations between adult salamanders and juveniles. We also present research on kin recognition, mutual mate guarding, social monogamy, and sexual coercion in *P. cinereus*.

7.1 INTERACTIONS OF ADULTS AND JUVENILES IN THE FOREST AND IN THE LABORATORY

As we showed in part 3.7, larger adults of *P. cinereus* of both sexes establish territories under and around cover objects on the forest floor (Mathis, 1989), whereas smaller adults act as "floaters" (Mathis, 1990a). *Plethodon cinereus* reaches sexual maturity only in its second or third year (Gillette, 2003) at Mountain Lake Biological Station (MLBS). Consequently, there is a long period of time during which juveniles interact with adults and with each other. Jaeger, Wicknick, et al. (1995) conducted a series of tests to determine (1) whether juveniles are excluded from or tolerated within the territories of adults, thus gaining access to prey-rich areas during dry periods, and (2) how juveniles respond to the territories of adults. In the forest at MLBS over an 11-day period during the noncourtship summer, Jaeger, Wicknick, et al. examined whether juveniles were consistently present in or absent from adult territories, or whether they entered those territories only during specific weather conditions (wet or dry). Each day they checked 100 natural cover objects along randomly selected transects from two 1 ha plots, recording the number, estimated age, and, in the case of adults, sex of the salamanders found under each cover object.

While adults were uniformly distributed among cover objects, the distribution of juveniles was clumped. Jaeger, Wicknick, et al. (1995) found 47 male–female pairs, 1 male–male pair, and zero female–female pairs. Only about 22% of the 257 juveniles were cohabiting with adults. There were no significant differences among the numbers of juveniles cohabiting with single adults of either sex or male–female pairs. However, juveniles were significantly more likely to be found under cover objects inhabited by adults during dry periods than during wet periods; over all 11 days, the association of juveniles with such cover objects was negative. This suggested that juveniles tend to move from the leaf litter into adult territories during dry periods and that juveniles are tolerated equally by both male and female adults, thus gaining access to prime foraging sites.

The tolerance of intruders into an adult's territory depends on the costs and benefits of sharing a territory. The cost of sharing limited food resources with juveniles could be low if juveniles ate different prey compared to adults. To determine if adults and juveniles share similar food sources, Jaeger, Wicknick, et al. (1995) examined the stomach contents of 403 salamanders (adults, 2-year-old and 1-year-old juveniles) from Shenandoah National Park (SNP). The number of prey items per salamander was significantly positively related to rainfall, and, during wet periods, the stomachs of juveniles contained more items than those of adults. Despite significant differences in width of prey items between adults and juveniles, food niche overlap values were generally high. The researchers inferred that while adults cannot leave their territories unprotected for prolonged periods of time in order to forage in the leaf litter, juveniles can maximize their energy intake by moving into the leaf litter during wet periods.

However, during dry periods, juveniles of *P. cinereus* would be forced to seek out moisture-retaining cover objects used by adults. In a laboratory experiment, Jaeger, Wicknick, et al. (1995) examined the reaction of juveniles to territorial pheromones of adults. Juveniles ($N = 28$) were offered the choice between two substrates (filter papers) in two randomized experimental and a control trials: (1) a male-marked versus a blank (water) substrate, (2) a female-marked versus a blank substrate, and (3) two blank substrates (control). Juveniles spent significantly more time on substrates marked by adult males and females than on blank substrates, indicating that they could detect territorial pheromones and were attracted to them. Jaeger, Wicknick, et al. inferred that juveniles of *P. cinereus* view adult territories as moisture and feeding refuges during dry conditions. They also examined the response of adults to juvenile intruders in the laboratory at MLBS. Territorial males ($N = 19$) were exposed, in randomized trials, to unfamiliar male intruders of a similar size and to 2-year-old juveniles. Males displayed significantly more all trunk raised (ATR) toward intruding adult males than toward juveniles. The authors inferred that adult male salamanders are more tolerant of unfamiliar intruding juveniles than of unfamiliar intruding adult males.

To examine the influence of familiarity on the interactions of adults with juveniles, Jaeger, Wicknick, et al. (1995) compared the behaviors of adults and juveniles that had previously cohabited in a territory in the forest with those of adults and juveniles that had been collected on different plots in the same forest. After 4 to 12 days of separation, the salamanders were able to distinguish between familiar and unfamiliar conspecifics. One-year-old juveniles moved toward previously cohabiting adults significantly more and showed significantly less EDGE behavior in the presence of familiar adults compared to same-age juveniles in the presence of unfamiliar adults. Also, territorial adults threatened previously cohabiting juveniles significantly less than juveniles with which they had not cohabited.

These results lead to a question: Why do adult salamanders tolerate juveniles in their territories? One hypothesis is that the juveniles that enter adult territories are kin (e.g., their offspring). Kin recognition can be tested by establishing molecular genetic relationships between cohabiting salamanders, as has been done at MLBS by Liebgold et al. (2006, in section 7.16). We discuss the potential for kin recognition by *P. cinereus* in section 7.18.

Research on interactions of adults and juveniles in the forest has focused on the distribution of salamanders under cover objects (but see section 7.3). Such moisture-retaining objects are important for survival of *P. cinereus* during dry periods. Mathis (1990a, in section 3.7) suggested that territorial quality may be determined by the size of the cover object. In a forest survey at MLBS, she observed that salamander size was positively related to cover-object size. Gabor (1995, in section 3.8), on the other hand, found no such relationship at SNP, but she did find that bigger salamanders had more food-rich territories. Therefore, Faragher and Jaeger (1997) further investigated the use of cover objects by salamanders by analyzing data on snout-vent-length (SVL) of salamanders and the sizes of their cover objects in a plot at SNP. They hypothesized that (1) on dry days, there should be a positive correlation between the SVL of the salamanders and the sizes of their cover objects, with a larger cover object assumed to provide a larger food supply; (2) on wet days, this correlation should be smaller, as salamanders should occasionally leave their defended territories and forage in the leaf litter; and (3) there should be no significant differences between the sizes of cover objects found with adults and those found with juveniles if adults do not exclude juveniles from their territories.

Faragher and Jaeger's (1997) data did not support the first two hypotheses, as salamander size was not significantly correlated with cover-object size on either wet or dry days. They suggested that as there were more salamanders than cover objects in this plot, there was competition for this resource. They inferred that body size might not be a true indicator of success in aggressive contests; alternatively, the size of the cover object may not be related to the size and quality of the available foraging area. Both of these inferences disagreed with Mathis's (1990a) conclusions from MLBS. Sizes of cover objects used by adults at SNP did not differ significantly from those used by juveniles; however, due to low statistical power, the hypothesis that adults are tolerant of juveniles entering their territories was only weakly supported by the data.

7.2 DISTRIBUTIONS OF ADULT MALES AND FEMALES

Understanding the spatial arrangements of adult salamanders on the forest floor is an important step in understanding the nature of intersexual associations in *P. cinereus* (Peterson et al., 2000). Mathis (1991b) had established that significantly more salamanders had neighbors of the opposite sex than of the same sex, and Jaeger, Wicknick, et al. (1995) found that adult salamanders cohabiting under a cover object outside the courtship season (summer) were almost exclusively male–female pairs; this was the first hint that socially monogamous relations not only occur but also endure during the noncourtship season. Jaeger et al. (2000) posed a model for alternative mating strategies based on the density of cover objects in the forest. This model was tested by Jaeger et al. (2001) and by Gollmann et al. (2004) at MLBS. They established two plots: (1) in an area kept free of natural cover objects into which 100 boards had been placed a year earlier (Jaeger et al., 2001) and (2) on a rocky slope (Gollmann et al., 2004). They searched for *P. cinereus* under these cover objects (boards and rocks, respectively) on 10 consecutive days during the summer noncourtship season.

Jaeger et al. (2001) found, in plot 1, that male–female pairs occurred significantly more frequently than expected by random chance and that single-sex pairs occurred less frequently than expected by chance (13 out of 37 females and 13 out of 23 males cohabited with a member of the opposite sex). The members of a pair were not always found together under the same board, but only the same individuals were recaptured under a given board. Jaeger et al. (2001) suggested that males and females of a pair make separate foraging excursions and then return to their specific cover object after feeding.

On rocky plot 2, with natural cover objects, Gollmann et al. (2004) found the same number of salamanders as in plot 1 with artificial cover objects, but they found little evidence for cohabitation of male–female pairs. However, several cover objects were used by both adults and juveniles, and two rocks were visited by more than one salamander of the same sex. Gollmann et al. (2004) suggested two tentative explanations for the differences in the spatial distribution of the salamanders in plots 1 and 2. First, the high density of cover objects in plot 2 led to a breakdown of territorial defense and a high proportion of nonterritorial floaters in the population (as modeled by Jaeger et al., 2000), or, second, the moveable cover objects in plot 2 were not identical to the territorial foci. Those foci might have been situated in cracks and crevices as well as under larger rocks that were too heavy to be turned (Gollmann et al., 2004).

Jaeger et al. (2000) suggested that the distribution of cover objects in a forest may influence the complex territorial system and mating strategies of *P. cinereus*. They hypothesized that

> H_1: In the case of widely spaced cover objects, salamanders could be forced to live either in socially monogamous pairs or to tolerate subordinate, same-sex-intruders.
>
> H_2: Groups of adjacent cover objects would allow the establishment of several nearby territories, with either males or females competing for the best territories and the members of the other sex establishing territories so as to encompass the territories of as many potential mating partners as possible.

This (H_2) would lead to various types of polygamy (polyandry, polygyny), depending on the number of individuals of each sex. Jaeger, Wicknick, et al. (1995, in section 7.1) had found, during the summer at MLBS, that about one-third of the salamanders were in male–female pairs (the "socially monogamous game") while two-thirds were "singles" (the "socially polygynous and polyandrous games").

> H_3: When the density of cover objects is very high, the benefits of territorial defense would break down, because the costs of defense would exceed the benefits, resulting in scramble competition for prey and mates.

On the forest floor at MLBS, cover objects vary in distributional density, resulting in what might appear as salamanders living as singles, in pairs, or as adults tolerating subordinate individuals (same-sex pairs), as found by Jaeger, Wicknick, et al.

7.3 MICRODISTRIBUTIONS OF ADULTS AND JUVENILES

While much research had focused on the distribution of salamanders under cover objects, Liebgold and Jaeger (2007) conducted an extensive survey on the spatial associations between juveniles and adults of *P. cinereus* both in the leaf litter on the forest floor and under cover objects at MLBS. They first examined whether the effects of atmospheric conditions on the activity of salamanders depended on the age class (with smaller individuals possibly more prone to dehydration), which might lead to a temporal separation. Second, they examined whether juveniles were nonrandomly distributed with respect to adults. If juveniles were positively associated with adults, this could be a result of kin discrimination or familiarity (Jaeger, Wicknick, et al., 1995), but if juveniles were negatively associated with adults, this could be because of competition or cannibalism. For intraspecific cannibalism of juveniles, see section 7.18.

Liebgold and Jaeger's (2007) results, from nocturnal surveys after rainfalls and cover object searches in late spring and summer on a forested plot at MLBS, did not support the hypothesis that 1-year-old juveniles and adults were differentially affected by meteorological conditions. They also found no evidence for temporal separation of the age classes. Measures of nearest-neighbor distances between juveniles and adults did not significantly differ from expectations of random chance. Therefore, they found no evidence for familiarity, kin discrimination, or competition. The distribution of juveniles did not appear to be strongly influenced by home ranges of adults, because they were neither positively nor negatively associated with adults during wet periods. There was a tendency, however, for juveniles to be located closer to adults than the radii of adult home ranges. This led the researchers to hypothesize that home ranges, in contrast to territorial foci, may not be strongly defended. For *P. cinereus*, we define "home range" as the area surrounding a cover object in which a salamander forages.

7.4 FEMALE–FEMALE INTERACTIONS

While salamanders found under the same cover object are frequently pairs of the opposite sex, Peterson et al. (2000) found about 22% of the females in same-sex pairs in the forest at MLBS during the autumn courtship season. This led Peterson et al. to investigate the frequency and nature of female–female associations in greater detail. They tested four hypotheses using both a forest and a laboratory study.

H_1: The observed values for female patterns of intrasexual pairing during the autumn courtship season are similar to those observed in the summer noncourtship season.

H_2: Males and females are equivalent in their propensity to form intrasexual pairs during the courtship season.

H_3: Females form strong affiliations with each other.

H_4: Single females preferentially move toward pheromones of unfamiliar males compared to those of unfamiliar females.

Peterson et al. (2000) were able to reject H_1 and H_2, because same-sex pairs were observed in the forest almost exclusively during the autumn courtship season. Of the two male–male pairs found, one pair was engaged in fighting when the cover object was lifted. During the summer noncourtship season, they found only two pairs of females brooding clutches (separated by >15 cm) under the same cover object. Although females were more likely to be found in pairs than males during the courtship season, the observed frequencies of same-sex pairs were lower than expected by random chance.

Using size-matched females in the laboratory, Peterson et al. (2000) compared the behavior of females toward their female partners (found together in the forest) with that of their behavior toward unfamiliar females. The focal animals were allowed to establish territories in Nunc bioassay dishes (chambers) for 5 days. On day 6, a salamander (unfamiliar or partner) restrained in a screen cage was placed on one side of the focal salamander's chamber and an empty screen cage on the other side. In a second trial, the positions of the salamander's cage and the empty cage were reversed. The focal females displayed no significant differences in aggressive or submissive behavior toward unfamiliar females compared to female partners. Peterson et al. inferred that the females found in pairs had no strong social affiliations with each other.

In another experiment, during the autumn courtship season, single gravid females were tested for their preferences of pheromones (on scented moss) from unfamiliar males or unfamiliar females. The focal females spent significantly more time on moss scented by females than by males, and nose tapped (NT) the moss with female pheromones more often than moss with male pheromones. Peterson et al. (2000) inferred that at least in some cases gravid females prefer to associate with unfamiliar females rather than with unfamiliar males during the courtship season.

Peterson et al. (2000) offered a number of hypotheses, which were not mutually exclusive, to explain female–female associations.

H_1: At the high densities ($2.8/m^2$) of *P. cinereus* at MLBS, adult salamanders may outnumber the number of suitable cover objects leading to the pairing of females under scarce cover objects.

H_2: Females may reduce their territorial aggression during the courtship season, leading to same-sex pairings.

H_3: Males tend to be more intolerant of unfamiliar intruders, including unfamiliar females, during the courtship season (Guffey et al., 1998, in section 7.9). Thus, given the limited number of territories available, females may prefer to associate temporarily with other females rather than with aggressive, unfamiliar males.

Peterson et al. concluded that further studies of male–female territorial associations are needed to better understand the complexity of social behavior in *P. cinereus*, which we review in sections 7.6 to 7.19.

7.5 MALE–FEMALE BEHAVIORAL INTERACTIONS IN THE FOREST

The spatial distribution of *P. cinereus* on the forest floor (in section 3.7) led to a number of inferences and hypotheses that were then tested in the laboratory. These laboratory studies led to the question of whether the behavior observed under laboratory conditions was an accurate representative of the behavior of salamanders in the forest.

Gergits and Jaeger (1990a) spent 10 nights during wet weather in the summer and autumn observing the interactions between salamanders in the forest at SNP. They followed each individual salamander for at least half an hour (or until they lost sight of it). Judging from their observations in the natural habitat, behavioral patterns in the laboratory were similar to those displayed under natural conditions. Looking toward (LT) another salamander was confirmed to be a threat behavior, as it caused one of the salamanders, usually the intruder, to move away. Gergits and Jaeger also observed focal salamanders (territorial owners) approaching an intruding salamander, biting it on the head, and then sometimes chasing the intruder as it moved away. Also, when salamanders approached a conspecific to touch it (NT), the other salamander reciprocated.

In autumn, they focused on courtship behavior. Of the 10 observed courtship interactions, 4 led to insemination. A detailed description of the males' courtship behavior can be found in Gergits and Jaeger (1990a), as follows. The male NT when crossing the presumed pheromonal trail of a female, followed it, approached the female in a submissive posture, and sometimes NT her body. The male always moved in front of the female, wriggling his tail. He rubbed his mental gland over the female's body until he reached her head and again wriggled his tail in front of her. The female rested her chin on the male's dorsum anterior to the vent; after having performed a "tail-straddling walk," the male deposited a spermatophore, which was then taken up by the female, whereupon the pair separated. Therefore, no postinsemination mate guarding was observed, but see section 7.11 for preinsemination mate guarding.

Other salamanders were attracted by the courting pairs; they LT them and in four cases disrupted the pair. Two of the courting males stopped to bite and chase the competing males that had approached the females, and two females walked away from the courting male after interference from another salamander. In one case, the female moved away after a male bit the courting male, and in the other case, the female left after a juvenile climbed over the tail-straddling pair. Thus the courtship behavior seen by other researchers in the laboratory was similar to what Gergits and Jaeger (1990a) observed on the forest floor.

7.6 THE ESS DATING GAME

During courtship, adults of *P. cinereus* may relax their territorial behavior. By allowing females to enter his territory, a male salamander could gain access to potential mates, while a female entering a male's territory may gain access to his sperm and to additional food sources. This led to two questions: How selective should a salamander be when allowing an opposite-sex intruder to enter its territory, and when is it beneficial to enter a territory as an intruder?

Female red-backed salamanders usually lay eggs only every other year at MLBS (Gillette, 2003, in section 7.19), because, presumably, the yolking of ova and the brooding of clutches require more fat reserves than females can acquire between adjacent breeding seasons. From a male's point of view, a nongravid female may be less attractive than a gravid one. By allowing a nongravid female to use the resources in his territory, a male does not gain immediate access to a mate; however, if that female stays with the male over a long period of time, then he could benefit by increasing his chance of mating with that now familiar female in the future.

In laboratory experiments, males defended their territories against both intruding males and nongravid females (e.g., Jaeger, 1984), but they were more tolerant of nongravid females than of males. Thomas et al. (1989) hypothesized that males would either act more aggressively toward nongravid females compared to gravid females or, alternatively, would exhibit no significant differences in their aggressive or submissive behaviors toward the two types of females. They allowed males of *P. cinereus* to establish territories in chambers containing an artificial burrow. They then introduced into a focal male's ($N = 29$) chamber either a gravid female, a nongravid female, a male, or a surrogate control (a rolled, damp paper towel) in random sequence and recorded the aggressive and submissive behaviors of the resident salamanders for 30 minutes. They found that resident males spent significantly more time threatening live intruders than surrogate controls. Bites were mostly directed toward male intruders ($N = 13$), whereas numbers of bites toward gravid ($N = 3$) and nongravid ($N = 1$) females were small. Differences in FLAT and EDGE behaviors among the four treatments were not significant. There was a slight tendency for resident males to be less submissive toward intruding gravid females than toward males and even less so toward nongravid females. In the control treatment, males spent significantly more time alone in a burrow than with a live intruder present in the experimental treatments.

So how does courtship occur when males are aggressive not only toward males but also toward female intruders? An a posteriori analysis of Thomas et al.'s (1989) data found that resident males became significantly more aggressive toward male intruders after having been exposed to gravid females, whereas their behavior toward nongravid females was not affected. The authors suggested that males may allow nongravid females to enter their territories in order to gain access to them for mating during the following year.

Hom et al. (1997) developed several evolutionary stable strategy (ESS) models to predict when males of *P. cinereus* should allow nongravid females access to their territories. Their models were based on the assumptions that males are territorial, that females reproduce biennially, and that males are either "nonpermissive" (excluding nongravid females from their territories) or "permissive" (allowing nongravid females to enter their territories). Hom et al. examined the conditions favorable for the maintenance and spread of permissive behavior in a population of *P. cinereus*. Based on these computer models, they concluded that there are two critical factors in the evolution of such permissive behavior: (1) female preference for permissive males and (2) the amount of food that males allow females to eat in their territories (see section 7.17) relative to the energetic costs of defense. Hom et al. inferred that females that are better at distinguishing between permissive and nonpermissive males may avoid aggressive contests with the latter. Females of *P. cinereus* would obtain more energy when allowed to feed in the permissive males' territories and thus have higher future fitness compared to females that lack the ability to distinguish between permissive and nonpermissive males. Hom et al.'s basic model suggested that permissiveness in males will evolve if females preferentially mate with permissive males. However, unless females mate exclusively with permissive males, nonpermissive males will also profit from increased fitness. Assuming that permissiveness is a heritable trait, this would provide a mechanism for maintenance of both permissive and nonpermissive behavior in a population of *P. cinereus*. Hom et al. termed their models "the ESS dating game."

7.7 MALES, FEMALES, AND FECES

A female entering the territory of a male might be faced with aggression or courtship behavior (as in section 7.6), and female mate choice may possibly be influenced by the quality of a male's territory (shown in section 3.8). An important indicator of territorial quality is the availability of nutrient-rich, easy-to-capture prey, such as termites, which *P. cinereus* prefers over ants (Gabor & Jaeger, 1995). Walls et al. (1989) hypothesized that females are able to assess the territorial quality of a potential mate without physical contact by examining fecal pellets and pheromones deposited by the territory's owner. They tested this hypothesis by first allowing females ($N = 24$) to establish territories in separate chambers with two artificial burrows and, after 1 day, placing a fecal pellet from the same male in front of each burrow: one it had produced after having fed on termites (high-quality diet) and one it had produced after having fed on ants (low-quality diet). The females did not spend significantly more time near either pellet, nor did the number of NT on or around the pellets differ. Females did spend significantly more time LT fecal pellets from males that had fed on termites than toward pellets from males that had fed on ants and significantly more time in the burrow near the fecal pellet from males that had fed on termites. Walls et al. inferred that females preferred to associate with males with a high-quality diet and inspected pellets visually (rather than olfactorily); Walls et al. (1989) could easily distinguish ant (black) from termite (white) pellets by their colors.

This conclusion was supported by a forest experiment that also included the examination of stomach contents of *P. cinereus* at SNP (termites have not been found by us at MLBS). Walls et al. (1989) found that males with termites in their diet had proportionally greater access to gravid females (most of which had also eaten termites) compared to males who had fed on ants. Although termites were rarer than ants, the 17% of males that had fed on termites had access to 47% of the potentially courting females. Thus females may profit from access to better food by selecting males with high-quality feeding territories, and those higher-quality males may benefit from better mating opportunities.

Walls et al. (1989) based their "sexy feces hypothesis" on the preference of gravid females for pellets advertising high-quality diets. However, they did not test the reactions of nongravid females and males to fecal pellets indicative of a high- or low-quality diet. Jaeger and Wise (1991) suggested that fecal pellets produced by a salamander that fed on termites might be interesting to conspecifics of either sex, because the fecal pellets might help an intruder assess the quality of a territory and thus whether it should engage in a territorial dispute. Jaeger and Wise tested the responses of intruding males, gravid females, and nongravid females ($N = 29$ for each type of intruder) to fecal pellets of resident males. Their results supported the sexy feces hypothesis. Females (both gravid and nongravid) spent significantly more time NT the substrate (or fecal pellets) than did male intruders; 16 gravid females, only 1 nongravid female, and zero males squashed fecal pellets (by pushing the snout into the pellet and rotating the head), presumably in order to obtain more chemical information about the resident's diet.

In a second experiment, Jaeger and Wise (1991) tested the responses of tailed and tailless male intruders toward the substrate and feces of tailed or tailless resident males. They found no significant differences in NT behavior (toward the substrate or pellets) between tailed or tailless intruding salamanders when confronted with tailed or tailless residents; none of the intruders squashed any pellets. Therefore it is unlikely that fecal pellets are involved in decisions of territorial contests by intruding males. The results of the experiments suggested, however, that fecal pellets may play a role in female mate choice. In the forest, females were more likely to be found near males with termites in their territories than near males with ants in their territories (Walls et al., 1989). Living on a nutrient-rich diet apparently can enhance a male's fitness in terms of gaining access to gravid females.

Prey species of *P. cinereus* range from soft springtails and termites to highly chitinous beetles and ants. Jaeger, Schwarz, et al. (1995) hypothesized that (1) males should prefer lightly armored prey, which should be easier to handle and digest than more chitinous prey types (as in Gabor and Jaeger, 1995), and (2) males need to learn less about foraging tactics effective in capturing lightly armored prey. For a period of three weeks, Jaeger, Schwarz, et al. fed males (N = 18 for each of the four treatments) just termites, just ants, a mixed diet containing termites and ants in equal numbers, or just *Drosophila*. Each male was then offered five ants and five termites simultaneously. They recorded the number of prey of each type eaten, the number of prey encountered within striking distance, the number of tongue strikes, the time until the first item of each type was eaten, and the time between encounter and ingestion.

Males from the different treatments did not differ significantly in the ratio of strikes toward ants versus termites or in the percentage of ingestions of termites (compared to ants). Most salamanders, however, ate more termites than ants. Even though more ants than termites came within striking distance, most salamanders (except for males trained on *Drosophila*) made fewer strikes toward ants than termites. The ratio of prey items within striking distance that were eaten was higher for termites, with no significant differences between treatments. The time to ingestion of the first prey item of each type varied significantly between treatments, with the shortest time for the salamanders trained on that specific prey type only. For ants, the ratio of ants ingested to the number of tongue strikes toward ants varied significantly with the most successful "anteaters" being salamanders that had been trained on just ants or ants and termites. Salamanders trained on ants also took significantly less time to ingest them. In the case of termites, capture success did not depend on training conditions. Based on the results, Jaeger, Schwarz, et al. (1995) concluded that (1) males recognize termites as superior prey and that (2) capturing ants efficiently requires learning. Thus the composition of fecal pellets provides useful information about an individual's foraging ability.

7.8 FEMALES PREFER LARGER MALES

The mating success of a male salamander may depend on his attractiveness or his territory's attractiveness as assessed by females (in terms of superior foraging abilities, possibly connected with "good genes" and high-quality food resources) and on his ability to defeat other males during direct competition for a female. In either case, body size (RHP) may play an important role.

Mathis (1991a) hypothesized that larger males are more likely to have access to females, either because of intrasexual competitive superiority or because of female discrimination. In a forest census on a 300 × 600 m plot at MLBS, she recorded the SVL of the salamanders found under rocks and logs and the positions of males relative to the positions of gravid females. Males that were encountered near gravid females were on average significantly larger than single males. These data supported the hypothesis that larger males have a greater likelihood of gaining access to females. Mathis then designed a series of laboratory experiments to determine if larger males were superior competitors or if females preferred larger males.

In the first experiment, females ($N = 26$) were allowed to establish territories for 5 days. On day 6, two restrained males ($N = 52$, in separate clear dishes with holes, so that the female could assess them both visually and chemically) differing in size were placed at opposite sides of a female's chamber, and the behavior of the female was recorded for 30 minutes. Overall, the time that females spent near either male did not differ significantly, but the proportion of females found near the larger male (42%) at the end of both trials was significantly greater than expected by chance. Females also spent significantly more time LT the larger male, NT more often in the vicinity of the larger male, and spent more time in EDGE when near the smaller male (Mathis, 1991a).

In a second experiment, three chambers were connected by tubes, with foam stoppers blocking the passages except at specific times. A female ($N = 24$) was placed into the central chamber and a male was placed in each of the end chambers. The SVL of these males ($N = 48$) differed by at least 5 mm. On days 2 to

4, each male was separately allowed access to the central chamber while the female was restrained as were males in the previous experiment. On day 5, both males were placed in the central chamber under habituation dishes for 15 minutes, and then the behavior of both males was recorded for 30 minutes. On day 6 and 7, each male was tested alone with the female, and his behavior was compared to his behavior when in the presence of the other male. Larger males spent significantly more time in ATR when the smaller male was present; otherwise there were no significant differences in their behavior. The behavior of smaller males was more affected by the presence of larger males. They spent less time near and less time LT the female. Smaller males displayed more submissive behavior (FLAT) and less threat behavior (ATR) in the presence of larger males. These results supported the hypotheses that (1) females prefer to associate with larger males and (2) larger males tend to dominate smaller males both when females are and are not present.

7.9 MALES AND FEMALES PREFER FAMILIAR OPPOSITE-SEX INDIVIDUALS

Mathis's (1991a) results suggested that larger males might gain more access to females because they are superior competitors and/or because of female preference. Because territorial quality was held constant in her experiments, female choice was not based on the resources of the territory held by a male but was influenced by characteristics of the male himself.

Under natural conditions, a male would be familiar with females that inhabit territories adjoining or overlapping his own (Mathis, 1991a) or with females that actually codefend the territory with him. Guffey et al. (1998) hypothesized that familiarity may lead to reduced aggression by either "dear enemy recognition" (Jaeger, 1981, in section 3.5) or long-term pair bonding (Gillette, Jaeger, et al., 2000, in section 7.10). Their experiment tested for the ability of *P. cinereus* to form intersexual bonds, as indicated by reduced aggression; this, in turn, might lead to a female being allowed to forage in a male's territory. Guffey et al. tested the behavior of adults of *P. cinereus* toward familiar and unfamiliar adults of the opposite sex, with special emphasis on the interactions of males with familiar and unfamiliar gravid females. Randomly assigned pairs ($N = 26$) of males and gravid females, collected during the autumn courtship season at MLBS, were allowed to establish territories in a test chamber, either together or separated by an airtight divider. They were fed on days 1 and 3, but not enough to become satiated. On day 5, the divider (if present) was removed and 5 to 10 termites were placed into the chamber. Guffey et al. recorded the proportion of termites eaten by each salamander and the behavior of both salamanders for 15 minutes. Controls consisted of termite-fed single males and single females that were alone with a surrogate.

Males threatened (ATR) unfamiliar females significantly more than familiar females or surrogates. They spent significantly more time in contact with familiar females and tended to show more submissive (FLAT) behavior toward them. Females were significantly more submissive (FLAT) in the presence of familiar males than with unfamiliar males. There were no significant differences in prey consumption by either males or females among the treatments. Guffey et al. (1998) concluded that adults of *P. cinereus* are less aggressive toward potentially courting individuals of the opposite sex when they are familiar with them.

The occurrence of male-female pairs in *P. cinereus* (Jaeger, Wicknick, et al., 1995, in section 7.1) could be explained by codefense of shared territories (section 7.11), intersexual overlap of territories, or visits of either males or females to the territories of members of the opposite sex. The question of which sex may be more tolerant of such intrusions is examined in section 7.19. Guffey et al. (1998) suggested that males are more likely to permit incursions by females. By allowing gravid females access to their territories, males can increase their chances of mating. Therefore, Guffey et al. predicted that males should allow as many females as possible to enter their territories as long as sufficient food resources are available. When food is limited, however, they should permit intrusion by a female only if the probability of mating with her is high (see section 7.17)

Apart from discriminating between familiar versus unfamiliar females, males might also assess the quality (future reproductive output) of a potential mate. For females, successful reproduction (in terms of clutch size) depends on their energy reserves (Wise, 1995, in section 3.9). Tail autotomy, which occurs during both antipredator defense and intraspecific social conflicts, can depress reproductive output, especially when it occurs during the courtship season. Meche and Jaeger (2002) hypothesized that males would preferentially associate with tail-intact females during the courtship season. In a forest survey at MLBS, however, during the spring and autumn courtship seasons, they found that a significantly higher proportion of females in pairs, relative to single females, had experienced recent tail loss. There was no such difference among males.

Meche and Jaeger (2002) designed a laboratory experiment that offered territorial males ($N = 25$, in a random sequence) a choice among (1) two gravid females of different sizes, (2) a larger, tail-autotomized gravid female and a smaller gravid female of the same total length (TL), and (3) two gravid females of the same size of which one had been induced to autotomize her tail (about 20 mm of it) 5 days prior to the experiment. Each male was allowed to establish a territory in a test chamber for 5 days. On day 6, two females that were restrained in separate transparent, perforated tubes were placed on opposite sides of the chamber. In the first treatment, males spent more time on the side with the smaller female, but in the other two treatments, males spent more time with the tail-intact females. Threat behavior by males occurred more often toward tail-intact females. The results did not support the hypothesis that males should prefer larger, tail-intact females. Females do prefer larger males over smaller ones (Mathis, 1991a), however, suggesting that females are choosier about their partners than are males.

Meche and Jaeger (2002) offered a number of alternative explanations for the high proportion of tail-autotomized females associating with males in the forest: (1) a preference of males for tailless females—which would have demonstrated that they could survive an attack—is not supported by the data collected in the laboratory, nor is it likely that (2) pairs will be more often attacked by predators than single salamanders, because there was no higher proportion of tail-autotomized male salamanders living in pairs. (3) Intruding male salamanders are more aggressive toward tailless as compared to tail-intact territorial males (Wise & Jaeger, 1998). If tail-autotomized females also encounter more aggression from intruders than tail-intact ones, they may prefer to align themselves with a male rather than defend a territory on their own. (4) Finally, and most likely, females' tail autotomy might be caused by their male partners, either through higher aggression at their first encounter (as in the previously discussed results) or in punishing socially polyandrous females (Jaeger, Gillette, et al., 2002, in section 7.12).

7.10 SOCIAL MONOGAMY

On the forest floor at MLBS, territories of salamanders overlap more or are shared more with those of opposite sex compared to same-sex adults (Jaeger, Wicknick, et al., 1995), and male–female pairs are frequently found under the same cover object both during the courtship and noncourtship seasons (Jaeger, 1979). These observations led to the hypothesis that adults of *P. cinereus* display behaviors characteristic of social monogamy. Gillette, Jaeger, et al. (2000) defined social monogamy as occurring "when two heterosexual adults, exclusive of kin-directed behaviour, direct significantly less aggression and significantly more submission towards each other, and/or spend significantly more time associating with each other relative to other adult heterosexual conspecifics" (1241). Staying together in socially monogamous pairs outside the courtship season may be beneficial to both partners, because females can profit from access to the resources in a male's territory and males may have a better chance of eventually mating with their female partners.

An important prerogative for social monogamy (and other complex social interactions) is the ability of the partners to recognize each other. For *P. cinereus*, this recognition can occur through chemical cues (pheromones) and/or visual cues (Kohn & Jaeger, 2009, in section 9.5). If salamanders can recognize their partners through chemical cues, then they should also be able to identify territories marked with pheromones by their partners. Gillette, Kolb, et al. (2000) hypothesized that (1) females collected in male–female pairs during the courtship season should be able to discriminate between substrate cues from partners and unfamiliar males, and (2) if such discrimination does occur, a female should direct preferential behavior toward her partner's substrate.

Females (N = 30) collected in male–female pairs during the autumn courtship season by Gillette, Kolb, et al. (2000) at MLBS were randomly subjected to four treatments. Female partners were given a choice between paired substrates in screen cages marked by (1) her partner versus a single male, (2) her partner versus an unmarked substrate, (3) a single male versus an unmarked substrate, and (4) two damp, unmarked substrates. The authors recorded the behavior of each focal female partner for 15 minutes and found that females spent significantly more time resting (front of trunk raised [FTR]) on the side of the chamber containing their partners' substrates relative to that of unfamiliar males. Females touched the screen cage containing the filter paper marked by their partners and NT near it significantly more often compared to the substrate scented by strangers. These results suggested that females prefer substrates marked by their partners over substrates marked by unfamiliar males. Females also displayed more aggressive behavior toward their partners' scents than toward the unscented controls.

Gillette, Kolb, et al. (2000) concluded that the partner need not actually be present to elicit preferential behavior, because chemical cues alone were sufficient. Chemical cues can help a female to avoid territories of unfamiliar males and also decrease the likelihood of her mating with males other than her partner (but see section 7.16 for evidence of polyandry at MLBS). The results of the study were consistent with *P. cinereus* engaging in social monogamy, at least during the courtship season. Social monogamy, however, does not necessarily imply mating monogamy (see section 10.4 for definitions). Possibly salamanders of either sex may engage in extra-pair copulations when a suitable opportunity arises (see sections 7.16, 7.17, and 10.4).

Finding that females can use chemical cues to recognize their partners led Gillette, Jaeger, et al. (2000) to examine if individuals in a pair could discern the pairing status of strangers and whether this would affect their behavior toward them. In a series of laboratory experiments, they compared the behavior of paired males and females (N = 29 pairs) toward their own partners versus unfamiliar single individuals or novel paired salamanders of the opposite sex. The results indicated that, during the courtship season, paired females were significantly attracted to their partners when the alternative was a novel paired male but not when it was a novel single male. They displayed significantly more ATR only toward paired, but not single, strangers. Alternatively, females might have responded to aggressive paired strangers but not to single strangers.

Paired males were attracted to their partners when the alternative was an unfamiliar paired female, but these males showed no such preference for their partners over a novel single female. Alternatively, they could have been repulsed from a paired but not a single unfamiliar female. Gillette, Jaeger, et al. (2000) found that salamanders could detect the pairing status of the novel salamanders, probably by the presence or absence of chemical cues on them derived from their partners. While the results are not inconsistent with the occurrence of social monogamy in *P. cinereus*, they do suggest that adult salamanders might employ a socially polygamous strategy when an unpaired individual of the opposite sex is available: that is, another variation of the ESS dating game modeled by Hom et al. (1997, in section 7.6).

In the control treatment, paired salamanders had a choice between their partners and a surrogate salamander; both males and females spent significantly more time on the side of the chamber with their (restrained) partner than with the surrogate. Also, salamanders touched their partners significantly more often and spent more time in FTR. These data are consistent with the idea that individuals in a pair are attracted to their partners rather than being repulsed by strangers.

During the noncourtship season, neither males nor females showed significant preferential behavior toward either their partners or unfamiliar paired adults of the opposite sex. Thus, these data do not support the hypothesis that socially monogamous associations continue beyond the courtship season (as found by Gillette, 2003, in section 7.19), nor do they preclude the possibility of pair associations lasting between courtship seasons (see Jaeger et al., 1995, in section 7.1, for pairs found in the forest during the noncourtship season).

According to Wittenberger and Tilson (1980), social monogamy "should" evolve when a territorial female pairing with an unmated male gains higher fitness values than this would gain by pairing with an already paired male or when aggression by mated females prevents males from acquiring additional mates, both of which (according to Gillette, Jaeger, et al., 2000) are applicable to red-backed salamanders at MLBS. Gillette, Jaeger, et al. did not support the hypothesis posed by Guffey et al. (1998) that salamanders defend separate territories but allow intersexual territorial intrusions. These two views led to two alternative explanations of the occurrence of male–female pairs. Either males and females jointly defend a single territory, or they maintain separate, overlapping territories, both of which may lead to the formation of long-term associations indicative of social monogamy.

7.11 MUTUAL MATE GUARDING

Lang and Jaeger (2000) addressed whether adults of *P. cinereus* codefend territories as female–male pairs. If codefense occurs, intruders should be expelled more efficiently by a pair than by a single territorial salamander. Lang and Jaeger tested the null hypothesis that paired adults (from MLBS and tested at the University of Louisiana at Lafayette [ULL]) codefending a territory would not significantly differ in aggressive or submissive behaviors toward territorial intruders of the same or the opposite sex during the courtship and noncourtship seasons. Alternatively, the salamanders might not forego future polyandrous and/or polygynous associations with single intruders of the opposite sex. They tested the prediction that paired individuals separated from their partners would be more submissive and less aggressive toward intruders of the opposite versus the same sex only during the courtship season.

Focal salamanders were allowed to establish territories in chambers for 5 days, either together as pairs or separated, during both the autumn courtship season ($N = 27$ pairs) and summer noncourtship season ($N = 29$ pairs). The future single intruders were kept in Petri dishes (which were inhospitable), each connected by a tube to an empty chamber so that they might learn how to escape from their Petri dishes. On day 6, each male or female intruder was allowed to enter the chamber of either a pair, just the pair's female, or just the pair's male. After the intruder had first entered the experimental chamber, Lang and Jaeger (2000) recorded aggressive (ATR, BITE), FLAT, and NT behaviors until the intruder left the chamber or until 15 minutes had elapsed.

The time that residents spent threatening intruders of either sex was not significantly different for pairs versus individuals separated from their partners, nor were there any significant differences between the time members of a pair spent threatening intruders during the courtship versus the noncourtship seasons. However, in pairs, males spent significantly more time threatening intruding males and females spent significantly more time threatening intruding females than did their respective partners both during the courtship and noncourtship seasons. Females separated from their partners during the courtship season spent significantly more time threatening intruding single females than did the males, whereas there was no such difference in males' and females' threat behavior toward intruding single males. Female residents that were separated or in a pair were more likely than males to bite intruders of either sex, both in the courtship and noncourtship seasons.

Intruding females spent significantly less time in territories defended by pairs versus separated adults of either sex during the courtship season. For male intruders, the differences were not significant. Intruders of both sexes spent significantly less time in territories defended by pairs during the courtship season than during the noncourtship season. Male intruders spent significantly more time touching separated males or females than paired ones. Female intruders spent significantly more time touching separated female residents than female residents in pairs but did not significantly differ in time spent touching separated or paired male residents.

Female intruders appeared to vary their aggressive behavior much more than did male intruders, depending on the season, sex, and paired status of the residents. In contrast to males, females spent significantly more time in ATR toward pairs in the courtship season compared to the noncourtship season. Females also spent more time threatening separated female residents than did male intruders; male and female intruders did not significantly differ in their threatening behavior toward separated resident males.

Lang and Jaeger's (2000) results indicate that male–female pairs of *P. cinereus* codefend territories but not in a cooperative way, because resident males and females living as pairs mostly threatened intruders of the same sex. This suggests that paired, socially monogamous residents are not willing to forego future chances of extra-pair mating: that is, trading social monogamy for social polygamy, as in section 7.17. However, this example of mutual mate guarding would reduce the opportunities for polyandry and polygyny: resident males attacked intruding males while resident females attacked intruding females when together as partners.

7.12 SEXUAL COERCION

Clutton-Brock and Parker (1995) proposed sexual coercion as a third type of sexual selection that occurs when (usually) males subject (usually) females to forced copulation, harassment, or intimidation. These coercions may occur even when females merely associate (not mate) with other males, that is, when they become socially polyandrous. Jaeger, Gillette, et al. (2002) tested their "intimidation" hypothesis that males of *P. cinereus* would show more aggression toward socially polyandrous (wearing another male's pheromones) female partners than toward socially monogamous female partners both during the courtship and noncourtship seasons. Natural pairs ($N = 26$ during the summer noncourtship season and $N = 24$ during the autumn courtship season) were collected at MLBS and tested in the laboratory at ULL. Female members of a pair were removed from their partner-shared chambers and placed in either an empty chamber for 5 days (control) or with an unfamiliar, single male for 5 days before being returned to their original partners. Jaeger, Gillette, et al. then recorded the behavioral interactions (TOUCH, ATR, FLAT, EDGE) between the reunited pairs for 15 minutes. In the noncourtship season, males spent significantly more time threatening socially polyandrous versus socially monogamous partners. Males also touched polyandrous partners significantly less often than monogamous partners and kept at a larger mean distance from them. During the courtship season, males displayed significantly more EDGE with polyandrous partners, kept at a larger mean distance from them, and bit them more often. The only significant difference in the behavior of females, irrespective of season, was an increased tendency to try to escape (EDGE) from their partners after they had been with another male.

These results are consistent with the hypothesis that male salamanders resort to sexual coercion (intimidation), but they could also be interpreted in another way. A male might respond aggressively just to another male's pheromones worn by his socially polyandrous partner. Jaeger, Gillette, et al. (2002) designed two further tests to support or refute this alternative explanation. Socially naïve (single) females were placed into empty chambers or into chambers containing a single male for 5 days before being introduced into the territory of an unfamiliar male. Both during the courtship and noncourtship seasons, males did not behave significantly differently toward these socially polyandrous versus socially monogamous unfamiliar females, nor did the two types of females show any significant differences in their behavior toward the males. Thus there was no evidence that the males mistook females wearing male pheromones for males.

Jaeger, Gillette, et al. (2002) suggested that sexual coercion is context dependent in terms of a male's previous investment of resources in a female. In dealing with their natural partners, males may have already invested in the future offspring of the females by sharing the food resources in their territories (and perhaps sperm) with them. When meeting an unfamiliar female, no such previous investment has taken place. The novel female might, however, become a future mate, irrespective of her previous associations with other males, with the new male accepting her. The females must choose between feeding in territories of more than one male or "staying home" and avoiding aggression from their partners. How the aggressive behavior of male partners affects the associations with their female partners still needs to be tested in the natural habitat.

In most scenarios of sexual coercion, females are regarded as passive partners (Clutton-Brock & Parker, 1995). In *P. cinereus*, however, a paired female has at least as much to lose through polygamy of her partner as do males, because she would have to compete for limited food resources with other females allowed to feed in her partner's territory. Given the aggression that females demonstrate toward intruders of the same sex (Lang & Jaeger, 2000, in section 7.11), Prosen et al. (2004) hypothesized that females would also use sexual coercion to punish ("intimidate") fickle partners.

Following similar procedures as in Jaeger, Gillette, et al. (2002), Prosen et al. (2004) tested for sexual coercion by female partners during the autumn and late spring courtship seasons. During late spring, females at MLBS end courtship and prepare to lay and brood their eggs. In the autumn, females were significantly more aggressive toward socially polygynous male partners compared both to socially monogamous male partners and socially polygynous male strangers. They spent less time touching polygynous partners and significantly more often NT them. In late spring, however, females displayed no significant differences in any of the recorded behaviors toward socially polygynous or monogamous partners. The response of females toward socially polygynous versus socially naïve strangers did not differ significantly in either season. During the late spring courtship season, socially polygynous male strangers spent significantly less time in ATR and TOUCH than did socially naïve strangers. It appears that recent social history may influence the behavior of a male toward a novel female. Thus social monogamy might be maintained by both males and females in two ways: first by expulsion of same-sex intruders from the codefended territory (Lang & Jaeger, 2000) and second by aggressively punishing socially polygamous partners (Jaeger, Gillette, et al., 2002; Prosen et al., 2004).

7.13 IMPERFECT INFORMATION DURING SEXUAL DISCRIMINATION?

In interactions between adult salamanders, correct identification of the other's sex is of great importance. Sexual identification in red-backed salamanders is based on visual and chemical cues (Kohn & Jaeger, 2009), although this information can sometimes be imperfect, as seen in chapter 3. Imperfect signals may cause misinterpretation of the signal by the receiver. However, multimodal signals (e.g., including sight and smell) are easier to detect, remember, and discriminate than unimodal signals (Rowe, 1999). Page and Jaeger (2004) investigated the role of information quality and sensory modality on sexual discrimination of intruders by resident salamanders. They (1) tested the response of male resident salamanders to conflicting visual and olfactory signals (intruding salamanders covered in secretions from the opposite sex) and (2) presented male resident salamanders with unimodal (chemical cues only) instead of bimodal (visual and chemical cues) signals.

In experiment 1, male salamanders from MLBS (N = 30) were allowed to set up territories in test chambers at ULL for 5 days. Each resident male was exposed, in random order, to (1) a surrogate control (salamander-sized piece of moist, rolled paper towel), (2) a live male, or (3) female intruder that had been swabbed with secretions from the opposite sex and an unswabbed (4) male or (5) female intruder. In experiment 2, cotton swabs containing secretions from either males or females, a mixture of both males and females, or a surrogate were used instead of live intruders. The surrogate in experiment 2 was a cotton swab moistened with spring water. Page and Jaeger (2004) recorded the resident salamanders' aggressive (ATR, BITE), escape (EDGE), and investigatory (TOUCH, NT) behaviors for 15 minutes.

Overall, the intruders in experiment 1 did not behave significantly differently for any of the behaviors recorded (full dataset). However, for intruders that were used only once (partial dataset), the analysis revealed a significant difference for time spent in ATR across treatments but no significant differences in pair-wise comparisons. Residents did not differ significantly in time spent in EDGE and number of NT, but they touched live salamanders of either sex significantly more often than surrogates (pairwise comparisons). Male residents spent more time threatening male intruders than intruding females or surrogates. Bites were directed at untreated ($N = 1$) and treated ($N = 3$) males as well as females ($N = 2$). In experiment 2, no significant differences in TOUCH, NT, and EDGE were detected. Males directed significantly more threatening behavior toward male-scented cotton swabs than toward either surrogates or female-scented cotton swabs. Comparisons between experiments showed that the males touched live females more often than the cotton swabs with female secretions. They also touched both treated males and treated females more often than cotton swabs with mixed secretions.

Page and Jaeger (2004) inferred that male residents are more aggressive toward male stimuli, independent of the number of sensory modalities. When presented with olfactory cues only, the aggressive response toward mixed signals was more similar to the one toward male secretions than toward female secretions. While olfactory signals alone were sufficient to stimulate aggression, additional visual signals were important for eliciting a touching response. Page and Jaeger's results provided little evidence that multimodal signals have superior discriminability in the identification of the sex of intruding conspecifics by males of *P. cinereus*.

7.14 RELATIONSHIP VALUE AND CONFLICT RESOLUTION

Aggressive behavior of a male or female toward a fickle partner might be tempered by the "value" that the relationship has for the individual. Prosen (2004) designed a series of experiments to test the applicability of de Waal and Aureli's (1997) "valuable relationship hypothesis" to male–female pairs of *P. cinereus*. The valuable relationship hypothesis was developed by de Waal and Aureli to explain postconflict resolution seen in monkeys and apes. A valuable relationship may develop between two individuals if, for example, they cooperate (e.g., mutual grooming) before a conflict arises. After such a conflict, two individuals that have developed a valuable relationship (e.g., cooperation) should "make up" faster or with less aggression than two individuals that have not previously developed a valuable relationship. Prosen suggested that this conflict resolution hypothesis might apply to male–female pairs of *P. cinereus* following a bout of sexual coercion (intimidation), as in section 7.12. One possible value of the relationship for *P. cinereus* could lie in joint defense of a territory against intruders of another species, in this case the confamilial species *Eurycea cirrigera*, at MLBS.

Pairs of *P. cinereus* ($N = 30$) were allowed to establish territories in test chambers for 5 days before an intruder (*Eurycea*) or a surrogate was introduced. Each member of the pair was also tested, in random order, separately in the same way. The threat behavior of each salamander in a pair was compared to each individual's behavior when interacting with the intruder alone. Males spent significantly less time threatening the intruder in the presence of their female partners than when they interacted with the intruder alone. There was no such difference when the intruder was a surrogate. Thus the value of the female to the male was in the amount of time or energy that he conserved by displaying less aggressive behavior when with his female partner. Females, however, showed no significant differences in their threat behavior in the presence of their male partners compared to when they were alone. This indicates an asymmetry in the relationship, because males may perceive females as valuable while females do not value males in the same way.

To ascertain whether males *recognized* the value of females, Prosen (2004) compared the responses of males toward socially polyandrous and socially monogamous female partners after an absence of 5 days, using the methods given in section 7.12. Those females had (1) previously had an opportunity to join the male in the defense against an intruding *Eurycea* and thus were presumably of value to the male ("potential benefactor") or (2) had not had such an opportunity ("nonbenefactor"). Males significantly increased their aggression toward fickle versus socially monogamous partners only in the case of returning nonbenefactor females. They spent significantly less time threatening (ATR) socially polyandrous females when they were potential benefactors compared to nonbenefactors. Therefore, males discriminated between potential benefactor and nonbenefactor female partners.

Next Prosen (2004) asked whether the female had to participate actively in the defense of a common territory in order to be perceived to be valuable by the male. He tested each male twice, in random order, with his female partner either restrained or free, when faced with an intruder. The aggressive behavior of males toward the intruding *Eurycea* did not differ significantly in the two trials. This suggested that the mere physical presence of the potential benefactor female, though restrained, was sufficient to influence her partner's aggressive behavior and thus be of value to him.

Although Prosen (2004) did not find any evidence that females value males, he also examined whether females would discriminate aggressively between socially polygynous partners (away with another female for 5 days) that had (benefactor) or had not (nonbenefactor) previously had the opportunity to join in territorial defense against an intruding *Eurycea*. There was no significant difference in female behavior toward the two types of partners. Prosen suggested that the failure of females to distinguish between the two types of partners may be because females did not value males in terms of territorial defense; that is, unlike males, females did not reduce their aggression toward an intruder while in the presence of their partner. A male may also value a socially monogamous relationship in terms of its potential paternity assurance. Females, on the other hand, may value access to a male's territory due to the quantity or quality of the food resources there (see section 3.8).

Prosen's (2004) experiments suggest that *P. cinereus* has evolved impressive cognitive skills (as further explored in chapter 9). Individuals appear to be able to formulate the value of a cooperative versus noncooperative partner and to respond accordingly when resolving a conflict (*sensu* de Waal & Aureli, 1997).

7.15 NATURAL VERSUS FORCED PARTNERSHIPS

Many of our studies on social interactions in *P. cinereus* were based on the assumptions that (1) males and females found in pairs on the forest floor have formed long-term associations with each other (e.g., Jaeger, Wicknick, et al., 1995, in section 7.1) and (2) they have chosen to cohabit a single territory (Gillette, 2003, in section 7.19). Prosen et al. (2006) examined two alternative hypotheses based on the relaxation of these assumptions.

H_1: Pairs collected in the forest might have been found under the same cover object by chance, in which case they might be familiar with each other but not share a common social history ("familiarity hypothesis"). Punishment of polygamous "partners" might be caused by their appearing unfamiliar because of strange pheromones, derived from other males or females, that they wear.

H_2: In the previous experiments, salamanders found under the same cover object were maintained together in the laboratory for weeks and might have developed a social relationship in the laboratory, not previously in the forest. Punishment of socially polygamous partners might occur when pairs had been merely housed together for extended periods of time in the laboratory ("captive housing hypothesis").

Prosen et al. (2006) collected single salamanders at MLBS and kept them together in chambers in forced male–female pairs for a short (5 days) or long (30 days) period. The experimental set-up was similar to the procedures used with natural partners (Jaeger, Gillette, et al., 2002; Prosen et al., 2004), with one set of tests with focal males and one set with focal females. Males showed no significant differences in any of the observed behavioral patterns toward socially polyandrous versus socially monogamous females in forced pairs. Females had a tendency to behave more aggressively toward socially monogamous versus socially polygynous partners in both short- (5 days) and long-term (30 days) association. In contrast to experiments with natural pairs, neither males nor females in forced pairs "punished" polygamous partners in any pairing. Prosen et al. concluded that naturally occurring pairs of *P. cinereus* may indeed be long-term associations between individuals that had developed their social history in the forest and that punishment of socially polygamous partners in such pairs is not the consequence of mere familiarity or laboratory methodology. This conclusion conforms well with Gillette's (2003) data from the forest at MLBS (in section 7.19).

7.16 FEMALES ARE OFTEN GENETICALLY POLYANDROUS

Aggression toward fickle partners is a possible mechanism for maintaining pair bonds in *P. cinereus*. However, social monogamy does not necessarily mean that either partner will forego opportunities for extra-pair matings (Lang & Jaeger, 2000, in section 7.11). Animals of many species that are socially monogamous are frequently not genetically monogamous. Liebgold et al. (2006) used molecular techniques to assess the paternity of clutches of *P. cinereus* collected at MLBS. Their objective was to determine if the rate of single-father clutches was higher in *P. cinereus* than in aggregate-breeding salamanders with scramble competition for mates. Furthermore, they hypothesized that a male attending a female and her eggs will have fathered the majority of the embryos in that clutch, as might occur in social monogamy.

Liebgold et al. (2006) collected entire clutches and tail clippings from females attending eggs and from four males found under the same cover objects as the clutches. They genotyped each individual at five microsatellite loci in order to assess parentage. Not surprisingly, the genotypes of the females were consistent with all the embryos in the clutches that they were attending.

Of the 13 clutches with more than two embryos, two had one sire, seven had two sires (one of which had sired the majority of offspring), and four had three sires. In the last case, none of the males involved had sired the majority of offspring. All four males found at nest sites were genetically compatible with at least some of the embryos in the clutches; on average, the attending males had sired about 38% of the embryos. Liebgold et al. concluded that polyandry is the predominant female mating system in *P. cinereus*. The average and maximum numbers of sires of a clutch were similar or slightly lower than in aggregate-breeding salamanders. Given the length of the breeding period (autumn plus spring at MLBS), a female should have ample opportunity for extra-pair matings, and she may form a pair with different males at different times (see section 7.17). Yet social monogamy by females may have a small fitness pay-off for males, because 15.4% of the clutches were sired by only one male. Recall that Jaeger, Wicknick, et al. (1995, in section 7.1), at this same site at MLBS, had found about 27% of individuals in apparently socially monogamous pairs with the other 73% apparently playing social polyandry and social polygyny tactics, as in the ESS dating game of Hom et al. (1997, in section 7.6).

7.17 SWITCHING FROM SOCIAL MONOGAMY TO SOCIAL POLYGAMY

The aforementioned studies concerning social monogamy by *P. cinereus* assumed that female–male pairs found together in the forest are partners that form long-term, preferential social associations (sections 7.1, 7.15, and 7.19). Studies in the laboratory confirmed that such partners behave toward each other in ways consistent with the hypothesis of social monogamy (section 7.10). However, this pleasant scenario was challenged by at least three studies. First, only about 27% of the salamanders at MLBS were found in male–female pairs in the forest during the noncourtship summer (section 7.1), suggesting that a majority of salamanders opted for social polygyny and social polyandry (section 7.16). Second, although mutual mate guarding was observed in the laboratory, both male and female partners were less aggressive toward unfamiliar opposite-sex intruders than toward same-sex intruders (section 7.11). This suggested that both members of a partnership are open to extra-pair associations and/or matings. Third, a large fraction of females at MLBS were found to be genetically polyandrous, bearing offspring, in the same clutch, from one to three males (section 7.16). The terms *social monogamy, mating monogamy*, and *genetic monogamy* are ambiguous, and we try to decipher this ambiguity in section 10.4. However, whatever *social monogamy* means when applied to *P. cinereus* (as in section 7.10), it is now clear that it does not necessarily lead to either mating or genetic monogamy in the forest at MLBS (section 7.16).

Wilcox (2006) experimentally tested *P. cinereus* from MLBS to gain a better understanding of seemingly socially monogamous partnerships. In experiment 1, she tested the hypothesis that females from natural partnerships have a faster rate of food ingestion than do nonpartner females in the partner males' territories. This had been one of the basic assumptions of how females might benefit in fitness from sharing territories with males as partners.

For treatment 1, Wilcox (2006) allowed natural pairs ($N = 28$) to establish territories for 4 days in separate chambers. On day 1, she fed each pair 65 flies (*D. melanogaster*), enough to satiate both partners, and removed excess flies after 2 hours. On test day 5, she again gave each pair 65 flies. She randomized the two treatments, in which the 28 males from treatment 1 were presented on day 1 with unfamiliar, nonpartner females in place of the partner females in treatment 2. They were fed 65 flies on day 1 and again on test day 5. During each test, Wilcox recorded for 15 minutes the number of flies eaten by each male and female and their times spent in ATR 2–5 (Fig. 3.1).

Males did not differ significantly ($p = 0.803$) in numbers of flies eaten between treatments, but partner females in treatment 1 ate significantly more flies than did nonpartner females in treatment 2 ($p = 0.0001$). These results supported the prior assumption that a female profits in a partnership by enhancing her foraging success when in the presence of a partner male.

Males spent significantly more time in ATR 2–5 when nonpartner females foraged in their territories compared to when their partners foraged there ($p = 0.0006$), and partner females were significantly less aggressive toward their paired males than were nonpartner females ($p = 0.003$). So again, as in section 7.10, both males and females exhibited preferential behavior (less ATR) toward partners compared to nonpartners.

Prey abundance may influence the social associations between males and females, because females of *P. cinereus* need abundant prey to yolk their ova and to prepare to survive long periods of brooding their eggs (see section 7.19). Males apparently require prey-rich territories that attract females as partners and/or mates (see Walls et al., 1989, in section 7.7). Therefore, Wilcox (2006) tested whether partner males switch from social monogamy to social polygyny (experiment 2) and partner females switch from social monogamy to social polyandry (experiment 3) when prey densities in their territories increase and vice versa. The rationale was that in prey-poor territories, hungry salamanders might act as socially monogamous partners such as to inhibit prey competition from intruders (as in mutual mate guarding; Lang & Jaeger, 2000, in section 7.11), while in prey-rich territories such competition would be reduced and partners could then explore extra-pair associations (social polygamy).

Experiment 2 tested the response of male partners in prey-rich and prey-poor territories. Each male (*N* = 28 from MLBS) was tested randomly in four treatments: with (1) his partner or (2) an unfamiliar female in a prey-rich territory (chamber) and with (3) his partner or (4) a different unfamiliar female in a prey-poor territory. Unfamiliar females had been size-matched (<2 mm difference in SVL) to the partner females to eliminate size-dependent differences in behaviors. Male and female partners and the unfamiliar females were separated for 5 days and trained to feed on 12 (prey-rich) or 2 (prey-poor) flies (*Drosophila*) per day; Jaeger (1980b) had found that adults need ~6.8 flies per day to maintain zero change in body mass. On test day 6, each male was reunited with his partner or found an unfamiliar female in his chamber, and each was given either two or 12 flies. Wilcox (2006) recorded for 15 minutes the time that each individual spent in ATR 2-5 under all four treatments.

Males in prey-poor treatments were significantly more threatening toward unfamiliar females compared to their partner females (typical of social monogamy), but in prey-rich territories, males exhibited no significant difference in threat (as if behaving in a socially polygynous way). In all treatments, unfamiliar females were significantly more threatening toward the males than were partner females.

Experiment 3 tested the responses of female partners (*N* = 30 from MLBS) in prey-rich and prey-poor territories toward partner and unfamiliar males. Otherwise, the methodology was identical to that in experiment 2. In prey-poor treatments, partner females were significantly more threatening toward unfamiliar males than toward partner males (as if socially monogamous) but not in prey-rich treatments (as if socially polyandrous). Males showed no significant differences in threat among treatments.

The results of these two experiments need to be confirmed or refuted by experiments in the forest, but such natural experiments would be extremely difficult to design. However, if males and females do switch back and forth between social monogamy and social polygamy in the forest, for whatever reasons, this might explain observations at MLBS. First, in section 7.1, only ~27% of the salamanders were found in pairs, with the others apparently living as singles. These tactics might reflect variance in prey densities among territories on the forest floor, as observed by Wilcox (2006) in the laboratory. Second, switching among social monogamy and singles might also explain why most, but not all, females at MLBS were found to be genetically polyandrous (as in section 7.16). Research by Gillette (2003, in section 7.19) in the forest at MLBS helps to clarify these issues.

7.18 BROODING BEHAVIOR AND NEONATES: KIN RECOGNITION?

Attending eggs is a full-time summer job for brooding females of *P. cinereus*. Female red-backed salamanders lay 2 to 14 eggs under rocks or inside rotting logs on the forest floor in June at MLBS and stay with them for 6 to 9 weeks until they hatch (Gillette, 2003), only occasionally leaving the nest site to forage. Other researchers have reported (reviewed by Peterson, 2000) that maternal care includes defense of the clutches against cannibalistic conspecifics, keeping in close contact with the eggs, thus inhibiting their desiccation and protecting them against fungal infection; that neonates stay with their mother for about two weeks until they have consumed their yolk reserves; and that brooding is costly for a female, because she lives almost exclusively on her fat reserves. Consequently, females at MLBS reproduce only biennially (but see section 7.19 for exceptions). The ability to home to their nest sites after displacement or to discriminate their own eggs from another female's eggs has been demonstrated for several species of plethodontid salamanders. In *P. cinereus*, however, egg recognition might not be under strong selection pressures, because females rarely leave their clutches and there are usually no other eggs in close proximity to a nest site.

Peterson (2000) tested, at MLBS, three alternative hypotheses concerning female discrimination of eggs in *P. cinereus*.

H_1: Egg recognition does not occur.

H_2: Females have direct egg recognition abilities.

H_3: Females use indirect mechanisms such as recognizing the nest rather than the eggs.

In a laboratory experiment, each brooding female ($N = 36$) was placed into a test chamber with two randomly assigned artificial burrows, one housing her own clutch and one with an unfamiliar clutch of about the same size. Most of the females stayed with one of the clutches during the entire 15-minute trial period. There were no significant differences in behaviors (time staying near a clutch, touching it, threatening it, cannibalism) directed toward their own versus another female's eggs. Also, all brooding females ($N = 10$) that were displaced 1 m from their clutches in a forest experiment at MLBS returned to their nests within 1 to 3 days. Peterson's (2000) data were more consistent with indirect mechanisms of egg recognition (H_3). She inferred that nest, not direct egg, discrimination occurs in *P. cinereus* and that nest-site recognition would be sufficient to avoid misdirection of parental care. Under natural conditions, females of *P. cinereus*, even if they should temporarily leave their nest site (e.g., in response to predation pressure), are unlikely to encounter unattended clutches that they might confuse with their own. Direct discrimination of eggs would only be of selective value in brooding aggregations of *P. cinereus*, as they have been observed in Michigan.

How does a female find her eggs after having been displaced? Peterson (2000) hypothesized that females use olfactory cues in "homing" to their nests. These olfactory cues could be from the eggs themselves or from females recognizing their own territorial scent marks at their nest sites. While such an indirect recognition of the clutch would serve to direct maternal care only toward a female's own eggs, this would not hold true for the period after hatching. Neonates stay with their mother for about two weeks (Peterson, 2000) and may be found under the same cover objects up to three months later (see section 7.19). Neonates may profit from staying in their mother's territory, because this would allow them access to food and defense by their mothers against predatory invertebrates (Jaeger & Forester, 1993).

In a series of laboratory experiments, Gibbons et al. (2003) examined mother–offspring discrimination. In experiment 1, they tested the hypothesis that mothers and their neonates (N = 33) would differ significantly in their behavior toward each other compared to unrelated individuals. Each female was tested, in random order, once with one of her own kin (from which she had been separated for 5 days) and once with an unrelated neonate. Neither the females nor the neonates showed any significant differences in behavior toward related compared to unrelated individuals. Experiments 2 and 3 tested the hypothesis that neonates could discriminate between olfactory cues from their mother compared to an unrelated female. In experiment 2, neonates (N = 27) were offered the choice between unscented moss and moss scented by either their mother or an unrelated female. In experiment 3, the neonates (N = 180) were allowed to choose between substrates with the chemical cues of their mother and an unrelated female. The neonates showed no significant preference for a specific moss or substrate in either of the experiments, though there was a slight trend toward maternal cues in experiment 3.

In experiment 4, Gibbons et al. (2003) tested the hypothesis that females defend their offspring against potentially cannibalistic adult intruders more than they defend unrelated neonates. Adult females (N = 24) were allowed to establish separate territories in test chambers for 5 days. Each resident female's behavior toward a female intruder was tested twice, in the presence of either her own or an unrelated neonate, in random order. During the 15-minute trials, females showed no significant differences in any of the recorded behavior patterns (ATR, FLAT, EDGE, TOUCH, NT) toward the female intruders. In experiment 5, the authors tested the hypothesis that females (N = 17) will cannibalize unrelated neonates significantly more often than their own offspring.

One related and one unrelated neonate were placed into the chamber of a hungry female. Twelve out of 17 females were cannibalistic, consuming both types of neonates. However, females significantly more often ate the unrelated neonate before their own neonate, thus indicating preferential cannibalization of nonkin neonates.

The results of this last experiment indicated that females of *P. cinereus* exhibit kin recognition but only discriminate kin in certain contexts (i.e., cannibalism). Given the low dispersal of the neonates (Gillette, 2003, in section 7.19), a female would be more likely to encounter her own offspring within her territory under natural conditions. The neonates were bitten and preyed upon by both unrelated females and their mothers. Gibbons et al. concluded that neonates may not recognize their mothers, but mothers respond favorably toward their neonates in the context of cannibalism.

Although evidence for direct kin recognition is weak in Gibbons et al. (2003), we propose an alternative hypothesis for indirect recognition. Neonates and older juveniles remain under or near their natal cover objects (as seen in section 7.19). They there become familiar to the residential adults, who respond favorably toward familiar juveniles compared to unfamiliar juveniles (as in section 7.1). Therefore, residential adults indirectly respond favorably toward their own offspring and unfavorably toward unfamiliar, nonkin offspring due to familiarity.

7.19 WHAT 3,487 UNIQUELY MARKED SALAMANDERS REVEAL ABOUT SOCIAL RELATIONSHIPS

Gillette (2003) spent three years residing near MLBS so that she could monitor the ecology of *P. cinereus* year-round at our research site there. Her goal was to track social interactions ("who lives with whom and where and when") and demographics of large and small adults, juveniles, neonates, and, if possible, females brooding their eggs. She especially wanted to test three hypotheses generated from previous studies in the forest and from behavioral experiments in the laboratory: (1) males compete intrasexually for feeding territories under cover objects (e.g., Gabor & Jaeger, 1995; Guffey et al., 1998); (2) females align themselves with territorial males (e.g., Gabor & Jaeger, 1995; Jaeger & Wise, 1991; Walls et al., 1989); and (3) adult males and females pair as socially monogamous partners, at least sometimes (e.g., Gillette, Jaeger, et al., 2000; Jaeger et al., 2001; Prosen et al., 2004; Wilcox, 2006).

Gillette (2003) began by clearing three well-separated plots of forest of all natural cover objects that *P. cinereus* could use as territories. In each of plots 1 and 2, she placed 100 small boards (30 × 30 × 2 cm) in 10 × 10 arrays, with boards spaced 4 m apart. In plot 3, she placed 200 boards in pairs such that each pair was 1 m apart with pairs separated by 4 m, and pairs in a 10 × 10 array. These boards were rapidly inhabited by the salamanders that, presumably, had been displaced from under their natural cover objects. Boards were staked into the ground to prevent bears, raccoons, and other predators from flipping them. Gillette then began to mark uniquely and later recapture all salamanders located under the boards, which summed to 3,487 salamanders (N = 1,933 females and 1,554 males) by the end of the three-year study. Boards were surveyed during daylight, but night surveys also located salamanders (marked or not) in the leaf litter between boards and outside of the plots, to locate nonterritorial (unmarked) floaters.

Gillette (2003) marked each salamander uniquely by injecting 2 to 6 color-coded dots, subcutaneously with elastomer dyes, into the translucent abdomen. To determine if these injections caused detrimental effects on behavior and/or survivorship, she conducted a controlled laboratory experiment (N = 48 salamanders) at MLBS. After 16 months in the laboratory, she tested for feeding efficiencies of injected and noninjected individuals and found no significant differences. Mortality was only 8.3% scattered between the treatments. She inferred that the injections had no long-term effects in the injected salamanders under boards in the forest.

Gillette (2003) recorded much data for each salamander when first found and subsequently recaptured in the forest, such as SVL, TL, tail condition (autotomized or intact), stripe (full length on dorsum, broken, or absent), and reproductive condition of females (with or without yolked ova as seen through the abdomen and mating status: mated recently with spermatophore cap in her cloaca or not). Each month she turned boards every 3 days when salamanders were not in winter retreats underground (they surface whenever the ground thaws even during winters). Her data after three years were voluminous, so we merely summarize the results here from her 31 tables and 14 figures of data and statistics.

7.19.1 Basic population ecology

Adult males were recaptured more frequently than females but over shorter time spans. These data supported the hypothesis that males are the primary territory holders; that is, they are more often recaptured under a given board but over shorter time spans presumably due to males competing with each other for territorial ownership. However, females were recaptured under the same boards over longer time spans, suggesting that they too are site tenacious and less likely to be displaced by other females.

Very few salamanders moved from one board to another board over the three years, even though *P. cinereus* can walk the 4 m distance in just a few minutes, as seen from previous homing experiments. Overall, Gillette (2003) recorded 3,095 recaptures of male salamanders and 3,397 recaptures of female salamanders. In only 1.9% of the cases, the salamander had moved from one board (where first marked) to another. Most movements were to adjacent boards (paired boards in plot 3 were treated as one board), and occasionally moving salamanders returned to their original boards. Thus both sexes were extremely site tenacious.

Note that 3,487 salamanders were marked under 400 boards (mean = 8.7 per board), but far fewer (zero to 3) were found under a given board on a given census day. Therefore, Gillette (2003) assumed that most salamanders were underground each day. Alternatively, some could have been "away" foraging in the leaf litter, or some could have been marked but seldom recaptured nonterritorial intruders. All three alternatives may have been applicable, especially on wet (rainfall) days.

Surprisingly, site tenacity in all ages of juveniles was similar to that of adults. Juveniles rarely moved from one board (where first marked) to another and even then just to neighboring boards. This refuted Gillette's (2003) and our assumption that juveniles are the dispersal stage, wandering the forest floor until reaching sexual maturity, when they would begin to quest for territorial ownership. Instead, juveniles stayed in their natal areas, suggesting that *P. cinereus* at MLBS cohabit in family groups, with relatedness quite high among salamanders in small areas (cover objects). Outbreeding is assumed, because males displace each other, as when for example younger, fully grown males displace older territorial residents. These data showing relatively short territorial tenure for males would fit this hypothesis. By this scenario, younger adults are the dispersal stage with males seeking territories and females seeking alignment with territorial males.

Of the 51 neonates that hatched and were marked under boards, only 1 moved to a neighboring board from August (hatch date) until November (when they disappeared underground for winter). None were found next spring under any board. They may have remained fossorial, died from winter's freeze, or been eaten by predators such as commonly seen large beetles and centipedes. However, these data imply that the highest mortality occurs at the small (>1 cm TL) neonate stage.

Some juveniles ($N = 30$) were too young to be sexed when first marked but later grew to sexual maturity. Juveniles later identified as males were recaptured more frequently than those later identified as females, as also seen in adults. Therefore, behavioral differences between the sexes begin at the juvenile stage (ages 1 to 2 or 3 years). This suggests that juvenile males gain more benefit than juvenile females from remaining under one cover object.

7.19.2 Size distribution and growth rates

The first-year age group grew to 32 mm SVL by the second September, which represents very rapid growth from neonate to age 13 months post-hatch. These juveniles were more often found under boards during summers, when mating does not occur, until September, when mating begins; then adults were more often under boards. This indicates that yearlings were displaced by territorial adults when mating began, which fits the previous hypothesis that adults spend considerable time foraging in the (when wet) leaf litter during summers. Growth rates from juvenile to adult stages did not differ between the sexes except for the largest adult size class (>45.0 mm SVL), when females continued to grow while males did not.

7.19.3 Sexual maturity and female fecundity

Matings occurred during both spring and autumn based on spermatophores seen in cloacae of females. Two females were observed to mated twice, suggesting that some females are polyandrous (as in section 7.16). Gillette (2003) suggested that multiple matings (if with different males) and the long mating period allow females to lessen aggression from males, but this would not conform to evidence of male aggression toward socially polyandrous females in section 7.12. Multiple matings by polyandry would increase genetic diversity among a female's offspring and, perhaps, allow for sperm competition and cryptic female choice (Eberhard, 1996).

Males reached sexual maturity at about 2 years of age (~37.0 mm SVL), based on the appearances of dark testes seen through the abdomen (Gillette & Peterson, 2001); females did not lay their first clutches until nearly 4 years old. Therefore, females probably reached sexual maturity near age 3 years when they mated.

Although females of *P. cinereus* generally brood clutches biennially, those that lost their clutches to predators early during brooding were able to lay another clutch the next year. Therefore, females that do not suffer the energetic costs of brooding are then able to forage (as seen by increased fat storage) and to yolk ova for the next year. This provides another opportunity for "permissive males" in the ESS dating game in section 7.6. Consecutive-year brooding by some females explains why more than half of the females in any given breeding season were gravid (i.e., carrying yolked ova).

These data also confirm that females face a trade-off by the energetic costs of brooding. Eggs do not survive without an attending female, but brooding successfully also inhibits a mother from foraging and thus from yolking ova for the next spring's laying.

Gillette (2003) also found some evidence for social monogamy (cf. section 7.10). Males and females often cohabited under boards during all seasons, and five of seven females known to have brooded eggs to hatching each had the same male with her more than once. These five males were sometimes in the nest cavities with the brooding females, and one male stayed with a female after she lost her clutch. Males did not assist in brooding, but Gillette suggested that attending males helped to defend nests from egg predators, especially intruding *P. cinereus*, possibly *Eurycea cirrigera*, and some small invertebrates (e.g. snails and ants).

7.19.4 Population size

Plot 1 contained 250 males and 388 females under 100 boards, allowing for one estimate of population size. This divides into 6.38 salamanders per board (possibly kin groups as described earlier) but only 0.49/m^2 in plot 1, far lower than Mathis's (1991b, in section 3.7) estimate of 2.8/m^2 at Gillette's (2003) site. However, population size varied dramatically from day to day, with increasing densities with increasing recent rainfalls.

Nighttime sampling in the leaf litter revealed a large population of individuals that did not appear under boards, namely, unmarked, nonterritorial, adult floaters of both sexes. However, contrary to a previous hypothesis, unmarked adult floaters and marked adult territorial individuals did not differ in SVL. Therefore, Gillette (2003) found no evidence that floaters were smaller and younger (and so less competitive in RHP) than individuals found under boards.

More females than males (i.e., 388 vs. 250 in plot 1) were found under boards. This supports the hypothesis that males compete for a limited number of cover objects and females align themselves to those territorial males. It also suggests how polyandry is achieved in territories, with females competing with each other for access to males' prey resources and sperm.

7.19.5 Sex ratio

The sex ratio was 1:1 female–male during springs and autumns but was mysteriously female-biased during summers, when about half of the females should have been brooding eggs underground (only a few nests occurred under boards). Gillette (2003) speculated that males might spend more time underground than females during summers, but she posed no plausible explanation as to why they should do so.

Sex ratios under natural cover objects outside of plots and in the leaf litter within plots did not differ from the ratios under boards. This indicated that the sex ratios under boards were representative of the above-ground population as a whole and that many salamanders of both sexes had failed to gain territories under cover objects.

7.19.6 Intersexual associational behavior

Gillette, Jaeger, et al. (2000) had found in the laboratory evidence for social monogamy by pairs found together at MLBS (see section 7.10); the results from Gillette's (2003) board censuses were consistent with the social monogamy hypothesis. The same female–male pairs were repeatedly found together under boards more frequently than expected by random chance. However, such pairs tended to remain together through only one reproductive cycle: from August when females began to yolk their ova until oviposition the following May or June, although some males remained with brooding females. These male–female pairings during the females' gravid months conform to Lang and Jaeger's (2000) observations of mutual mate guarding in section 7.11. Why then did Jaeger, Wicknick, et al. (1995) find only ~27% of the population in female–male pairs during the summer (section 7.1) near Gillette's three plots? Perhaps males and not yet gravid females begin to explore new partnerships during summers, as in the ESS dating game of section 7.6, while the remaining females are brooding their clutches.

Gillette (2003) found no evidence for social polygamy, because males and females did not trade partners under the boards. "No individual, however, was ever seen more than once with more than one individual of the opposite sex" (102), she wrote. This suggests that frequent mating/genetic polyandry (as found in section 7.16) occurs by short-term extra-pair matings by intruding, floater females and males that briefly gain access to the paired males and females. Gillette suggested that sexual coercion by both sexes (as in section 7.12) may drive the social system toward social monogamy in the forest. This would not necessarily prevent brief extra-pair matings; a partner cannot always be mate-guarded while away foraging, and the cost of being bitten during bouts of sexual coercion may be offset by enhanced fitness from extra-pair matings, especially for males.

Some evidence found a connection between female mating frequency and pairing status. Females seen associating with the same males at least three times were more likely to mate (spermatophores in their cloacae) than females not found multiple times with one male. Therefore, social monogamy may lead to enhanced mating by females but not necessarily with her partner male.

Gillette (2003) concluded that some, but not all, individuals of *P. cinereus* at MLBS form long-term (9–10 months) intersexual associations (social monogamy). Thus far, this is unknown for other species of Amphibia.

7.19.7 Consequences of tail autotomy

When a salamander autotomizes its tail, the tail regrows but the stripe on the previously lost portion of the tail does not. Most individuals of *P. cinereus* at MLBS have stripes, so Gillette (2003) could detect both recent and historical tail autotomies and their consequences for allocation of energy by males and females over 3 years. Historical tail loss affected growth rates in adult males but not in females, such that males directed surplus energy, beyond that required for maintenance, into growth more than did females. Growth rates in juveniles were not affected by tail loss. Recent tail loss, however, did affect a female's reproductive output, because considerable fat is stored in the tail. Recent tail autotomies reduced females' clutch sizes, yet females with a small portion of tail loss increased clutch size relative to those with no recent tail loss. Females with 70% to 100% tail autotomy suffered reduced clutch sizes. Gillette inferred that adult males allocate more energy into growth while females allocate more energy into egg production.

This inference would conform to previous hypotheses that females prefer to align with larger males (in sections 7.7 and 7.8) and larger males can better defend territories (in sections 3.7 and 3.8). Gillette (2003) also suggested that substantial recent tail loss presents a trade-off for females: they put less energy into yolking ova (smaller clutch size) so as to use their remaining fat reserves for overwintering survival or maximize clutch size and risk mortality.

7.20 A PRELIMINARY MODEL OF SOCIAL ORGANIZATION WITHIN *P. CINEREUS*

Gillette's (2003) heroic research shows that 3 years of intensive data collection in the forest is still insufficient time to gain a complete understanding of the social relationships within *P. cinereus*. Yet her survey allows us to integrate her data with those of other forest and laboratory studies in chapters 3, 4, 5, and 7. However, our current tentative view of social organization applies only to the population at MLBS, because Wise and Jaeger (2016, in section 3.10) found that territorial behavior of *P. cinereus* varies among geographic localities, and so social organization may vary as well.

Neonates and juveniles do not disperse but remain at or near their natal sites. Growth at these life-history stages is rapid, so males reach sexual maturity in ~2 years and females in ~3 years post-hatch. Adults tolerate juveniles under their cover objects because they recognize them either as kin or as just familiar cohabitants. Females appear to recognize their own offspring in that mothers favor cannibalization of other mothers' neonates rather than their own neonates;

such presumed "kin recognition" by males has not yet been tested. However, adults displace juveniles from home sites during the courtship seasons.

Once juveniles reach sexual maturity, they probably disperse away from their home sites, if only for short distances, when both adult males and females begin to seek cover objects as territories. These young adults remain as floaters for years until they grow large (or experienced) enough to gain and to defend territories. As floaters, they remain in the leaf litter but go underground as the litter dries.

Cover objects vary in the quantity or quality of prey found under or near them, and larger males obtain territories containing better sources of prey (e.g., termites vs. ants). Females prefer to align themselves with larger territorial males (cued by the prey quality in males' feces), but such males have a relatively short tenure of territorial ownership due to male–male competition for prey-rich territories. Females have longer tenure in territories because they are better at resisting female intruders. Inbreeding is perhaps minimized by the turnover of males under cover objects. Once a territory has been established for at least five days, the male and/or female defend it aggressively by threat displays and biting while scent-marking the territory with pheromones that contain a plethora of social information. Territories allow the occupants to forage by the rules of "optimal foraging theories" except when defending against intruders.

When a female and male cohabit a territory, they frequently, but not always, form a socially monogamous relationship, especially when prey are scarce there, with mutual mate guarding. These relationships tend to last for only one reproductive cycle, until the female lays her eggs, at which time the male explores new relationships with other females that are just beginning to yolk their ova for the next spring's laying. However, socially monogamous females, and presumably males, may engage in extra-pair matings leading to genetic polyandry and clutches with two or more fathers. Social (if not mating) monogamy is driven by mutual sexual coercion (intimidation via threat displays and biting) toward partners that even associate with other-sex individuals. Yet prior territorial cooperation against intruders leads to less aggression toward a partner that has been in an extra-pair association.

Once territories have been established, both females and males are reluctant to leave them, except during foraging trips into the leaf litter after rainfalls, while defending them aggressively against same-sex intruders. Such aggression can lead to fights resulting in tail autotomy or damage to the nasolabial grooves, both losses to future fitness for the loser of the contest. Tail autotomy results in males putting more energy into continued growth while females suffer smaller clutch sizes, partly due to the high energetic cost of brooding. Social monogamy, with codefense of territories, appears to lead to stability in the distribution of territories on the forest floor, depending on the distribution of cover objects; that is, the territorial behavior breaks down when cover objects are closely spaced such that territorial intrusions by neighbors plus floaters are frequent.

Obviously this scenario poses many as yet untested hypotheses. We propose this preliminary view of social organization merely to stimulate future research by other behavioral ecologists. Also, based on previous laboratory experiments, we expect considerable variation among conspecifics (see section 10.1) in their choices of social organizations (e.g., to be or not to be socially monogamous as partners within territories).

Interlude – The research summarized in chapters 3, 4, 5, and 7 suggest that *P. cinereus* can alter its behaviors based on learning and on a suite of ambient information. Some examples are (1) judging prey densities in a given area while optimally foraging, based on some type of numerical rule; (2) displacing aggression between conspecifics, which is an expression of "anger" (*sensu* Darwin, 1872); and (3) learning to distinguish between unfamiliar and familiar conspecifics when expressing aggression. Chapter 9 examines some aspects of the cognitive abilities of *P. cinereus* that underlie its complex decision-making processes, but first, in chapter 8, we examine how red-backed salamanders and ringneck snakes engage in predator–prey "games" (*sensu* Maynard Smith, 1982).

7.21 SELECTED RECENT RESEARCH BY OTHERS: SOCIAL BEHAVIOR

Much research has been performed concerning how chemical cues influence social behavior. Jaeger and Wise (1991), following Walls et al. (1989), proposed that female *P. cinereus* use fecal inspection and squashing behavior to evaluate male territorial quality and, potentially, as a form of mate assessment. Karuzas et al. (2004) studied an alternative hypothesis for fecal inspection and squashing behavior by *P. cinereus.* They exposed gravid females maintained on high- and low-quality diets to fecal pellets of both male and female conspecifics fed high-quality diets. They found that females that fed on low-quality diets showed similar rates of squashing and inspection of male and female feces. They proposed that these results indicate that females use fecal pellets to select foraging areas. They also inferred that males may exploit female fecal inspection behavior as a way that attracts foraging females to their territories, because this would increase the male encounter rate with females.

In section 10.1, we note that there are assortative associations in *P. cinereus.* Red-backed salamanders are polymorphic, with two common color phenotypes: striped and unstriped. Anthony et al. (2008) found evidence for assortative pairing of striped and unstriped phenotypes in the forest in Ohio. Following this research, Acord et al. (2013) studied assortative "mating" in two populations using natural cover objects. They found that striped females paired with striped males were significantly larger in body size than those paired with unstriped males. Following this, they performed laboratory studies to determine the methods by which assortative mating occurred. They found no evidence that chemical cues or diet cues from feces of males contributed to assortative mating. However, females of both phenotypes were more likely to associate with striped males. These results suggest that striped males are more attractive to females, hence their associational patterns in the forest.

From the work presented in this chapter, it is clear that social interactions have strong effects on the behavior of *P. cinereus*. Kinship and familiarity (also see chapter 9) may play a role in these social interactions. Because *P. cinereus* has low levels of dispersal and its density in the forest of Virginia is high, it is likely that individuals interact with both familiar and related individuals. Liebgold and Cabe (2008) found that juveniles of *P. cinereus* housed with familiar adults grew at higher rates than those housed with unfamiliar adults during one year of a 2-year study. Relatedness, on the other hand, did not affect growth. Thus while *P. cinereus* may be capable of kin recognition (Gibbons et al., 2003), this ability may not play a role in future fitness benefits for juveniles.

Not only does familiarity affect social interactions, but Liebgold (2014) also found that the social environment modifies future juvenile behaviors. He found that the behavior of adults from two different areas (Virginia vs. Michigan) differed in territorial behavior and aggressiveness and that these differences affected the social behavior of juvenile salamanders in future interactions with other juveniles. Specifically, juveniles that were housed with more aggressive or more territorial adults showed more EDGE behavior when interacting with other juveniles, suggesting that they were acting like "losers" or trying to "avoid" conspecifics. These results indicate that social interactions and environments can affect future fitness consequences.

Lang and Jaeger (2000) found that male–female pairs of *P. cinereus* codefend territories against members of the same sex. They concluded that some level of mutual mate guarding occurs for *P. cinereus*. In a study with two species in the *Eurycea bislineata* complex, *E. aquatica* and *E. cirrigera*, Deitloff et al. (2014) found that neither species defended territories. However, they did find that males of *E. aquatica* performed mate-guarding behavior more so than did males of *E. cirrigera*.

We now turn to chapter 8, where we explore the many ways in which red-backed salamanders interact with snake predators.

8

Predator–prey interactions between *P. cinereus* and a snake

Predation pressures on terrestrial salamanders are poorly understood, because seeing predators and prey interact under leaf litter and in fossorial burrows is nearly impossible. Some laboratory studies have been conducted, but these are of little value, because they mask the salamanders' tactics of predator avoidance on the heterogeneous forest floor (see sections 6.4 and 6.5).

At Mountain Lake Biological Station (MLBS), a few likely predators on *P. cinereus* are obvious. Shrews may be the primary predator of adults; we have observed shrew burrows under cover objects where *P. cinereus* is usually missing there. Some species of snakes (*Thamnophis*) probably eat salamanders near streams, but they seldom move far from streams into the upland forest. Brown bears turn cover objects (such as our artificial ones; Gillette, 2003), but *P. cinereus* would be a tiny morsel for a large bear and so of little value in its diet. Forest birds (e.g., thrashers, towhees, and thrushes) could be significant predators by turning leaf litter, but Jaeger (1981a) observed that these birds are only minor predators on *P. cinereus*; they are efficient at turning dry leaves (when *P. cinereus* is absent there) but inefficient at turning wet leaves (when salamanders are present there). Therefore, we suggest that adults of *P. cinereus* at MLBS are largely resistant to predation, which may account for their large density at MLBS (mean = $2.8/m^2$; Mathis, 1991b). We have estimated that the maximum lifespan for *P. cinereus* is 25 to 30 years at MLBS. Most predation on *P. cinereus* probably occurs at the egg, neonate, and young juvenile stages. We have observed eggs being eaten by ants and snails at MLBS, and brooding females cannot defend their eggs against such oophagy. Also, neonates and small juveniles (<2 years old) may be eaten by centipedes and large coleopterans. Therefore, the heaviest predation pressures for *P. cinereus* almost certainly occur during the first two years of ontogeny. One forest-dwelling species is known to be a major predator on *P. cinereus: Diadophis punctatus*, the ringneck snake (Petranka, 1998). This species is absent from our high-elevation research sites at Hawksbill Mountain, Shenandoah National Park, and MLBS, but it and *P. cinereus* are broadly sympatric at lower elevations.

Lancaster (1994) chose to test the hypotheses that *P. cinereus* can, by olfaction and vision, detect the presence of *D. punctatus* and that the snake can, by olfaction, detect the presence of the salamander. She suspected that these two species may engage in a "coevolutionary arms race" (Dawkins & Krebs, 1979) in which the salamander is under selection pressure to detect and avoid the snake and the snake is under selection pressure to detect and capture the salamander. Theoretically, selection should be more intense on *P. cinereus* than on *D. punctatus* due to the "life–dinner principle" (Dawkins & Krebs, 1979): if a predatory attack is successful, the salamander will lose its life (and future fitness), but if the predatory attack fails, the snake will lose only a meal. Lancaster collected salamanders and snakes from Beaver Island and Garden Island, in Michigan, because both species are abundant there and *D. punctatus* feeds almost exclusively on *P. cinereus* in Michigan. Her nine experiments were conducted in the laboratory at the University of Louisiana at Lafayette, and we briefly summarize these behavioral experiments in sequence.

8.1 CAN *P. CINEREUS* DETECT THE SNAKE VISUALLY OR CHEMICALLY?

Experiments 1, 2, and 3 tested the hypotheses that *P. cinereus* can detect the presence of *D. punctatus* by chemical and/or visual cues. Snakes were allowed five days to leave presumed chemical odors on paper towels, and strips of these were used as presumed chemical cues for the salamanders. A test chamber contained either a blank strip control on each side or a snake-marked strip randomly placed on one side with a control on the other side. Each salamander was tested randomly once in each condition for 30 minutes, and data recorded were time spent on each side of the chamber, in EDGE behavior, immobile (an antipredator behavior), and number of nose taps (NT). Lancaster (1994) tested territorial resident salamanders ($N = 22$ in experiment 1) and nonresident salamanders ($N = 22$ in experiment 2) separately. The results were not significant for any recorded behavior.

For experiment 3, Lancaster (1994) reasoned that *P. cinereus* might need both chemical and visual cues to detect the nearness of the snake. Again, presumed chemical cues were obtained from the snakes, but the snakes were then placed in transparent glass Petri dishes for the visual cues. Empty or surrogate-occupied Petri dishes served as controls in the appropriate conditions. Each salamander was tested randomly in four conditions: (1) chemical cue versus control, (2) visual cue versus control, (3) both cues, and (4) neither cue as the general control. For 20 minutes Lancaster recorded times immobile and in EDGE and number of nose taps. Again the results were not significant for any behavior. Lancaster inferred that either (1) *P. cinereus* is incapable of identifying the predator's chemical (or visual) cues or (2) *D. punctatus* does not place chemical information on the substrate while foraging. We have found that *D. punctatus* does produce strong cloacal odors when handled by us (which might support option 1), but option 2 might be an evolutionary adaptation for a foraging snake that must search for and "sneak up on" its prey.

8.2 CAN THE SNAKE DETECT CHEMICAL CUES FROM *P. CINEREUS*?

In experiments 4 and 5, Lancaster (1994) tested the response of the snake to substrate odorants from *P. cinereus*; from chapter 5, Lancaster knew that *P. cinereus* does leave such odorants (scent marks) on substrates as well as volatile odorants in the air. She tested the responses of the snake to towels marked by *P. cinereus* on one side of a chamber versus a control towel on the other side in experiment 4 ($N = 10$ snakes) and to towels marked by a nonprey salamander (*Ambystoma opacum*) versus control in experiment 5 ($N = 12$ snakes). In both experiments, a general control was used with control towels on both sides of the chamber. The control towels had been marked by only *Drosophila virilis* because the salamanders had been fed this fly while marking their towels. A pilot study had shown that the snake readily ate *P. cinereus* but ate zero *A. opacum*. Each snake was placed into a narrow tunnel leading into the test chamber and allowed 15 minutes to habituate there before the door was opened into the test chamber. The primary response variable recorded for 30 minutes was the total time that each snake spent on each side of the experimental chamber compared to the general control.

Lancaster (1994) found that *D. punctatus* spent significantly more time on the side of the chamber containing odors of *P. cinereus* than in the control ($p < 0.0017$) but not significantly different times on substrates marked by *A. opacum* versus the control ($p > 0.05$). Similar results were found for two other of the snake's foraging behaviors, for example, less time moving while on the salamander's odor versus controls. Lancaster inferred that *D. punctatus* is "capable of detecting and identifying substrate chemicals originating from their prey" (70).

8.3 NAÏVE SNAKES RECOGNIZE ODORS OF *P. CINEREUS*

In experiment 6 Lancaster (1994) tested whether detection of chemical cues from *P. cinereus* is innate for *D. punctatus.* In experiments 4 and 5, she had tested adult snakes that probably had many prior experiences related to finding and eating *P. cinereus* in Michigan. Here she tested if prey-naïve snakes can also detect that salamander's chemical cues. She had collected gravid female snakes in Michigan, let them oviposit in the laboratory, and then tested the never-fed neonates ($N = 25$) randomly in two conditions: swab rubbed (1) on *P. cinereus* and (2) on the cricket *Acheta domesticus*, which is not a natural prey for *D. punctatus* (as she confirmed in a pilot study). She monitored the neonates' number of tongue flicks (a chemodetection behavior) toward each type of swab. The prey-naïve neonates directed significantly more tongue flicks toward the swabs with chemical cues from *P. cinereus* than toward cricket-control swabs. She inferred that the prey-naïve neonates were able to discriminate the salamander's substrate odors without prior learning and did so even though the adult salamanders were too large for the neonates to ingest.

At this point, one might suppose that *D. punctatus* has won the evolutionary arms race with *P. cinereus.* The snake (both adults and neonates) can detect the presence of pheromones from adult *P. cinereus* but the salamander appears unable to detect the presence of this snake by either its chemical or visual cues. However, as experiments 7 and 8 indicate, the salamander may have more subtle methods of inhibiting capture by *D. punctatus.*

8.4 TAIL AUTOTOMY DECEIVES THE SNAKE

Adults of *P. cinereus* will readily autotomize their tails when attacked by a predator (and by collectors who carelessly handle this species); a snake will then be attracted to the still-wiggling tail while the still-alive salamander departs from the "scene of the crime." In experiments 7 and 8, Lancaster (1994) hypothesized that *P. cinereus* will produce, from its tail, olfactory cues that are more attractive to *D. punctatus* than olfactory cues from the rest of its body.

In experiment 7, she tested, in random sequence, adult snakes' ($N = 11$) response to two conditions: (1) tail odors versus cricket controls and (2) body odors versus cricket controls. Tail odors were derived from swabbing the ventral side of a salamander's tail and body odors were derived from swabbing the ventral side of its abdomen. Cricket controls came from rubbing similarly the ventral side of *Acheta domesticus*. In the experiment, Lancaster (1994) observed for 60 seconds the rate of tongue flicks (RTF) and latency to begin RTF for each snake in each condition. She found that the snakes directed significantly more RTF toward odors from the tails than toward odors from the abdomen ($p < 0.01$), but there were no significant differences in latency to RTF. In experiment 8, she tested the responses of adult snakes to substrates marked by (1) tailed salamanders versus fly control and (2) tailless salamanders versus fly control. She had induced tail autotomization with forceps to create tailless salamanders. The tailed salamanders were sham-treated by autotomizing just the tips of their tails.

The results showed no significant differences for RTF, but the snakes spent significantly more time on the side of the chamber with odors from tailed salamanders than the side with odors from tailless ones ($p = 0.016$). Lancaster (1994) inferred that *P. cinereus* (though conspicuous to foraging *D. punctatus*) chemically diverts an attack by the snake toward its tail. In another study, where live salamanders confronted snakes in chambers, she saw 55.6% of snake attacks directed toward the tail, and 20% of these resulted in tail autotomy leaving the salamander otherwise unharmed. If this simple laboratory observation is a reflection of survival from attacks in the heterogeneous leaf litter habitat, then survival (with or without a tail) would be 54% to 62% of all snake attacks, depending on interpretations of the data. This research concerning the odors of tail-autotomizing salamanders was later published in much more detail by Lancaster and Wise (1996).

8.5 THE SNAKE FOLLOWS THE TRAIL OF *P. CINEREUS*

In experiment 9 Lancaster (1994) asked if *D. punctatus* can locate (trail) a salamander after detecting its pheromones. In a large chamber containing soil and a dividing panel, she placed one *P. cinereus* and required it to walk around the edge of half of the chamber and one *Ambystoma opacum* (the salamander control) and required it to walk around the edge of the other half of the chamber. She then removed the salamanders and the dividing panel and placed one snake ($N = 16$) in the chamber's unmarked center. For 5 minutes, she observed the snake's times (1) on each side of the chamber and (2) following each of the two trails. She statistically compared each of these times against a random expectation of time. She found that *D. punctatus* spent significantly more time on the side marked by *P. cinereus* than on the side marked by *A. opacum* ($p = 0.043$) and also significantly more time following the trail of *P. cinereus* than that of *A. opacum* ($p < 0.0003$). Both times were significantly greater than expected by random movement in the chamber. She inferred that *D. punctatus* is capable of following the trail of wandering *P. cinereus*.

8.6 THE PREDATOR–PREY EVOLUTIONARY ARMS RACE

For a broader perspective, we summarize in Table 8.1 our understanding of the evolutionary arms race between *D. punctatus* and *P. cinereus*. It appears from Table 8.1 that the arms race is either (1) equally balanced between the predator and its prey or, depending on how one interprets the predator–prey interactions, (2) *P. cinereus* has a slight advantage, as predicted by the life–dinner principle (Dawkins & Krebs, 1979). In either case, the antipredatory behavior of *P. cinereus* is almost exclusively defensive while the predator's behavior is active detection, pursuit, and attempted capture of its prey (Table 8.1).

Lancaster (1994) noted that *P. cinereus* becomes immobile (usually in the FLAT posture) once it is startled and that this is an effective defensive behavior

Table 8.1 SUMMARY OF A POSSIBLE EVOLUTIONARY ARMS RACE IN THE PREDATOR–PREY RELATIONSHIP OF DP AND PC

Advantage to DP	**Advantage to PC**
Sympatric with PC at lower elevations	Allopatric from DP at higher elevations
Shares the same forest floor habitat with PC	Better knowledge of its microhabitat (e.g., location of the escape hole)
Can detect pheromones of PC even at the prey-naïve neonate stage	Directs attention of DP toward its tail's pheromones
Can follow the trail of pheromones left by PC	Can leave confusing trails while actively searching for prey (e.g., crossing-over of the trail)
Can strike at PC	Can avoid strike by remaining immobile once startled
Can successfully capture the body of PC	Can avoid capture if strike is to its autotomizing tail
PC cannot detect odors or visual cues of DP	If grasped by DP will secrete copious slime secretions on its skin, making it difficult to hold
PC is not noxious/toxic if ingested	Its slimy secretions can inhibit the ingestion abilities of DP
If PC is not captured by a first strike, DP can pursue it rapidly for a second strike	If not captured by a first strike, PC can move rapidly to its escape hole

NOTE. DP = *Diadophis punctatus*; PC = *Plethodon cinereus*.

against *D. punctatus* (see Table 8.1). The snake is inhibited from striking at a motionless prey. She suggested that an immobile salamander confronting a waiting snake may lead to a "war of attrition" (Maynard Smith, 1982) between them. In this war, if the salamander moves first, the snake can strike, but if the snake moves first and gives up waiting, the salamander will be safe, at least for the moment. Profitable future research might focus on which animal has the longer giving-up time: the threatened salamander or the hungry snake?

8.7 SELECTED RECENT RESEARCH BY OTHERS: PREDATOR–PREY ARMS RACES

Following Lancaster's (1994) pioneering work, other scientists have explored predator–prey interactions using *P. cinereus*. A nice set of studies from Dean Adams' laboratory (Iowa State University) has been performed exploring Batesian mimicry by *P. cinereus*. As noted in chapters 7 and 10, red-backed salamanders are polymorphic; the two most common morphs are the striped and unstriped morphotypes, which are not considered conspicuous. In some localities where *P. cinereus* is sympatric with red-spotted newts, *Notophthalmus viridescens*, such as in western Massachusetts, a third red-orange morph (erythristic) is found, and its coloration appears similar to that of *N. viridescens*. Kraemer and Adams (2014) studied an interesting predator–prey arms race in a mimicry system between the model, *N. viridescens*, and its Batesian mimic, the erythristic color morph of *P. cinereus*. The red-orange coloration of toxic *N. viridescens* likely signals a conspicuous warning signal to predators of its unpalatability. *Plethodon cinereus* is not noxious and has a geographic distribution that overlaps broadly with *N. viridescens*. Kraemer and Adams found that mimics are only found with models, indicating that the presence of unpalatable models at a locality is necessary. Moreover, they demonstrated that birds select for mimicry on the basis of coloration but not brightness. Following this study, Kraemer et al. (2015) found that erythristic *P. cinereus* (the mimic) more closely resembled local *N. viridescens* (the model) than *N. viridescens* found at other localities. These results provide support for the hypothesis that selection for mimicry drives mimics to resemble local models. They also demonstrated that the *P. cinereus* mimics are less conspicuous than their models, indicating that predator–prey arms races are driving these differences and, currently, *P. cinereus* has minimized the costs of mimicry by matching model coloration while being less conspicuous to potential predators.

While Lancaster (1994) did not find evidence for *P. cinereus* using chemical or visual cues to detect the presence of *D. punctatus*, others have found evidence for the use of chemical cues in predator recognition. Indeed, antipredator response of plethodontid salamanders, and *P. cinereus* specifically, using chemical cues from predators has been well studied (reviewed by Madison et al., 2002). Red-backed salamanders show antipredator responses to cues directly released by the predator. For example, *P. cinereus* avoided substrates with body rinses from the garter snake, *Thamnophis sirtalis* (Madison et al., 1999). *Plethodon cinereus* also used alarm cues while detecting predators, as salamanders avoided areas with rinses from injured conspecifics and from *D. ochrophaeus* (Sullivan et al., 2003). One hypothesis is that avoiding these areas is an adaptive mechanism for evading areas with high predation risk. Further, *P. cinereus* used diet cues to recognize predators because they avoided areas with cues from predators that had recently been fed conspecific prey (Madison et al., 1999). These and many other studies have revealed that *P. cinereus*, like many other animals, can recognize predators using numerous types of chemical cues, each important for an antipredator response in different scenarios.

We now turn to chapter 9, where we examine aspects of the cognitive abilities of *P. cinereus* that allow for decision-making in a complex environment.

9

Cognitive ecology of *P. cinereus*

The term *cognitive ecology* was coined by Real (1993) for studies that combine ideas from cognitive science with those from behavioral ecology. Cognitive ecology focuses on the effects of information processing and decision-making on animal fitness (Dukas, 1998). In this chapter, we describe research on numerical discrimination, displacement behavior, learning, and individual recognition memory in *P. cinereus* within an ecological context.

9.1 NUMERICAL DISCRIMINATION BY *P. CINEREUS*

According to Stephens and Krebs (1986), animals should evolve foraging strategies that maximize their net energy gain when foraging (i.e., the energetic profit when foraging should exceed the energetic loss during foraging). Thus animals given a choice between two small numbers of food items should "go for more." Previous research with *P. cinereus* found that salamanders employ an optimal foraging strategy in that they forage indiscriminately between two sizes of fruit flies when both are in low numbers but specialize on the larger flies when the numbers of prey increase (Jaeger & Barnard, 1981, in section 4.3), and *P. cinereus* has the ability to change foraging tactics, which suggests that a salamander can assess the number of prey items within its visual field (Jaeger, Barnard, et al., 1982, in section 4.5). This led Uller et al. (2003) to investigate the ability of adults of *P. cinereus* to "go for more" (choose the larger of two numerosities) in a series of spontaneous forced-choice discrimination experiments using numbers of fruit flies, *Drosophila virilis*.

Salamanders were collected at Mountain Lake Biological Station (MLBS) and tested at the University of Louisiana at Lafayette (ULL). Uller et al. (2003) conducted 3 days of pretesting prior to the first day of the experiment. On day 1, they placed individuals into Nunc bioassay chambers lined with moist paper towels and each salamander was fed 5 live flies. On day 2, they placed a T-shaped testing enclosure into the test chamber (Fig. 9.1). They removed all uneaten flies, placed a transparent plastic tube (45 mm long with a 5 mm diameter) on each side of the large end of the enclosure and placed the salamander into the T-shaped chamber so that it could move around. On day 3, the salamander was again placed into the T-shaped chamber and was allowed to move around in the T. Day 4 was the test day. Two hours before the test, the researchers confined each salamander to the narrow tunnel part of the T and blocked it off with a piece of blue-tinted plastic (3.5 × 1.25 cm). Five minutes before the test, they

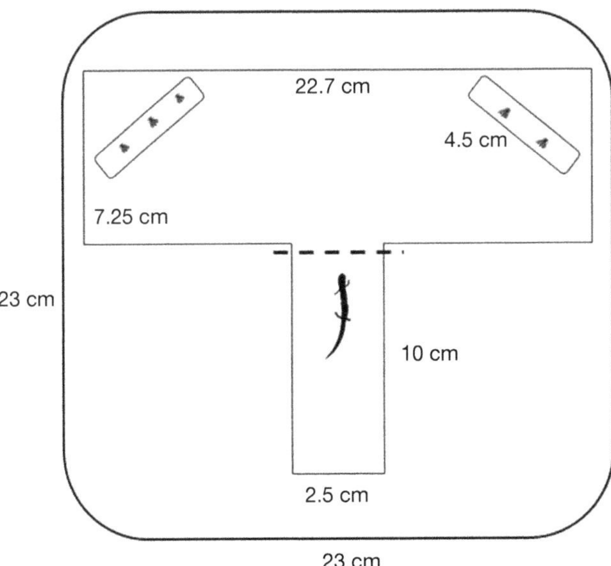

Figure 9.1 The chamber used in the numerical discrimination experiments, redrawn from Uller et al. (2003). The salamander is shown in the tunnel, blocked from entering the T-shaped chamber by a door. The tube on the left contains three live fruit flies while the tube on the right contains two flies, as in experiment 1. Drawing by Nancy Kohn.

removed the empty plastic tubes from the chamber and replaced them with two identical tubes containing x (e.g., 2) or y (e.g., 3) live fruit flies (Fig. 9.1). The tubes were placed 20 cm apart and equidistant from the salamander's path of approach from the tunnel. The tubes were sealed so that neither flies nor fly odors could escape. The test began when the salamander exited the tunnel, and the test ended when (1) its snout touched one of the tubes or (2) the salamander did not move or touch a tube before the cutoff time. They recorded the time that each salamander took to make a choice and measured an individual's choice for one numerosity or the other. Also, the side of each numerosity (x, y) or (y, x) was counterbalanced across subjects. Salamanders were used only once, and each experiment ended when 30 salamanders had made a choice.

In experiment 1, Uller et al. (2003) tested male discrimination of 2 versus 3 flies following similar experiments on rhesus monkeys, *Macaca mulatta* (Hauser et al., 2000) and human infants (Feigenson et al., 2002), which showed that these organisms can discriminate between these two numerosities. Twenty male salamanders approached and touched the tube with 3 fruit flies over the tube with 2 flies, and 10 touched the 2-fly tube ($p = 0.049$). Thus, based on a one-tailed binomial test, salamanders significantly chose 3 flies over 2. This experiment was followed by a second one with a few changes: in experiment 2, a narrow beam of light was pointed at the salamanders to stimulate them to leave the tunnel and move toward the flies. Twenty-one males touched the 3-fly tube and nine touched the 2-fly tube ($p = 0.022$). In experiment 3, the researchers found that females also significantly preferred the 3-fly tube over the 2-fly tube ($p = 0.049$). Overall, these results supported the numerosity hypothesis in which both male and female salamanders "go for more."

Discrimination is an animal's cognitive ability to recognize the number of objects encountered without needing to count them. Given that the discrimination limit is 4 for both monkeys and human infants (Feigenson et al., 2002; Hauser et al., 2000), experiment 4 tested a male forced-choice discrimination test of 4 versus 6 flies. Sixteen males touched the 6-fly tube and 14 touched the 4-fly tube ($p = 0.429$), indicating that the selection for fly number was random. Experiment 5 repeated this test with a new set of salamanders and found the same results ($p = 0.181$). In experiment 6, Uller et al. (2003) tested female discrimination of 3 versus 4 flies and again found that selection for fly number was random ($p = 0.429$). In experiment 7, they found that females touched a 2-fly tube significantly more often than a 1-fly tube ($p = 0.009$). Therefore, salamanders have a capacity to choose the larger numerosity in a spontaneous forced-choice condition of 1, 2, or 3 fruit flies. These results set the upper limit on numerosity at 3, which is similar to human infants (Feigenson et al., 2002).

Salamanders selected the larger of two numerosities when the numbers paired were 1 versus 2 and 2 versus 3, but not 3 versus 4 or 4 versus 6. Therefore, salamanders can recognize that 2 is more than 1 and 3 is more than 2. Thus salamanders go for more when the numerosities are smaller than 4. Uller et al. (2003) inferred that the salamanders spontaneously tracked the number of fruit flies because no training was involved and thus no learning. These results support the hypothesis that *P. cinereus* can discriminate among small numbers of prey, as suggested in section 4.5.

Ruby (2004) tested the hypothesis that spatial distance between or among flies influences the numerical discrimination of prey by *P. cinereus* from MLBS. Salamanders could either be attracted to a large grouping of prey because of the size of their mass or perceive flies as individual prey items. He designed four experiments in which live flies (*D. virilis*) were either grouped together in the tubes, as in Uller et al. (2003), or were spaced apart in separate subcompartments within the tubes. Using the same experimental design as in Uller et al., he found that the salamanders ($N = 44$ per experiment) responded significantly more to a spaced-out distribution. These results suggested that *P. cinereus* does distinguish among the number of visible prey items and not just among the size of a closely grouped mass of prey, which might appear as just one large prey item.

9.2 BOTH LEARNING AND HERITABILITY AFFECT FORAGING ABILITY

Sections 4.3 (on optimal foraging) and 9.1 (on numerical discrimination of prey numerosities) suggested that *P. cinereus* can adjust its foraging tactics according to the quality and abundance of available prey. This led Gibbons et al. (2005) to investigate the effects of learning (section 4.6) and heritability on foraging behavior in *P. cinereus*. They hypothesized the following.

> H_1: That these salamanders would improve their foraging efficiency with additional exposure to the same novel prey, regardless of age (neonates, yearling, and adults).
>
> H_2: That there would be a detectable prehatching (e.g., maternal or genetic) component to foraging behavior in this species, such that there would be significant clutch differences and heritability estimates for foraging traits and learning rates.

They also tested three secondary hypotheses:

> H_1: That clutch differences would be significant across all behaviors monitored.

H_2: That neonate behavior would be significantly correlated with yearling behavior (i.e., a neonate's behavior is a good predictor of that salamander's behavior as a yearling).

H_3: That neonates that survived to be tested as yearlings would be more efficient foragers as neonates than neonates that did not survive to the yearling stage.

Gibbons et al. (2005) collected 51 females with their eggs near MLBS and tested the salamanders in the laboratory at ULL. In the laboratory, females were placed with their clutches into separate Petri dishes lined with moist filter paper and a 10-cm piece of PVC pipe set on its side that acted as a burrow. They fed the females *D. virilis* until the first egg hatched then not again until all the eggs were hatched and the neonates were removed from their mothers. Each neonate was placed in its own small Petri dish lined with a moist filter paper.

In experiment 1, Gibbons et al. (2005) exposed ($N = 175$) neonates from 29 clutches to novel prey, termites (*Reticulitermes flavipes*). All neonates were fed as two-week-olds and had not been fed prior to the test. Researchers removed the paper towel from the neonate's Petri dish 5 minutes before the trial. Then they placed four termites into the center of the Petri dish, and during a 10-minute trial they recorded three behaviors: (1) snap, when a neonate snapped at, but did not capture, a termite; (2) capture; and (3) escape, when the salamander had the termite in its mouth but did not consume it. They measured (1) time to first capture, (2) intercapture intervals, (3) proportion of prey items captured, and (4) accuracy of the strike at prey. Each neonate completed three identical trials separated by 12 to 16 days.

In experiment 2, Gibbons et al. (2005) exposed yearlings ($N = 136$ from 51 clutches) to novel prey, *D. virilis*. The protocol was similar to experiment 1 except (1) yearlings were fed termites once a week until the experiment started, (2) the order of testing yearlings was randomized, (3) yearling Petri dishes did not have paper towels for cover, (4) the yearlings were fed four flies during the test, and (5) the three trials were separated by 7 to 19 days. In experiment 3, they exposed mothers ($N = 51$) to novel prey (termites). They used the same protocol as in experiment 2.

Gibbons et al. (2005) found that neonates showed a significant decrease in time to first capture with increased exposure to termites: that is, they learned to forage efficiently. Time to first capture for trial 1 was significantly longer than both trials 2 and 3. Trial 2 was also significantly longer than trial 3. Neonates showed a significant clutch difference in time to first capture over the three trials, suggesting a heritable component to capture time. For intercapture interval, trial 1 was significantly longer from both trials 2 and 3, but trials 2 and 3 were not significantly different from each other. Neonates captured significantly fewer prey in trial 1 than in either trials 2 or 3, and trials 2 and 3 were not significantly different. Neonate accuracy in trial 1 was significantly lower than in both trials 2 and 3. Trials 2 and 3 were not significantly different. Phenotypic correlations revealed that the proportion of prey captured was significantly negatively correlated with time to first capture and intercapture interval. The first capture was negatively correlated with accuracy.

For yearlings, survival analyses yielded a significant feeding trial, clutch effect, and time to first capture. The time to first capture for trial 1 was significantly longer than for both trials 2 and 3. There was no significant difference between trials 2 and 3. There were similar results for intercapture interval. For the proportion of items captured by yearlings, there was a significant feeding trial and clutch effect but no significant interaction between these effects. Yearling performance in trial 1 was significantly worse than in trials 2 and 3, but trials 2 and 3 were not significantly different. Yearlings were significantly less accurate in trial 1 than in trial 2, but there was no significant difference between trials 1 and 3. Yearling accuracy was significantly higher in trial 2 than 3, and all four traits were significantly correlated.

Adult females decreased their time to first capture with increasing exposure to the same novel prey (also see section 4.1). Time to first capture was significantly longer, the proportion of prey items captured was significantly lower, and intercapture interval was significantly longer in trial 1 than in both trials 2 and 3. Trials 2 and 3 were not significantly different from each other for any of these variables. Time to first capture was negatively correlated with proportion of prey items captured.

Neonates that did not survive to yearling age captured significantly fewer prey items during the three trials of experiment 1 than did neonates that survived to be tested as yearlings. Neonate sibling analyses found a significant heritability estimate for time to first capture for the second feeding trial and for intercapture interval for all three trials. There were also significant heritabilities for all feeding trials for proportion of prey items captured by neonates, including learning rate.

Gibbons et al. (2005) made the following inferences: (1) *P. cinereus* of all ages shows an increase in foraging efficiency with increased exposure to the same novel prey, because learning occurs after the first encounter with prey; (2) *P. cinereus* of all ages shows ingestional neophobia as they are less likely to consume a novel prey item during the initial exposure than they are during subsequent exposures; (3) there is a heritable component to foraging behavior in this species as data indicated that rapidity of learning may also be genetically based.

Salamanders that survived to yearling age were better foragers than salamanders that died before experiment 2, because survivors captured more prey in experiment 1 than did nonsurvivors. The results suggest that even though learning is an essential and a rapid component of foraging behavior, genetic effects also may influence variation between clutches, because there was a significant clutch difference across all behaviors monitored for neonates but not for yearlings. Thus both learning and heritability play a role in foraging behavior in *P. cinereus*.

9.3 DISPLACEMENT OF TERRITORIAL AGGRESSION

While sections 9.1 and 9.2 examine the cognitive ecology of foraging behavior, the next three sections focus on how learning and memory are important in various types of social interactions. Previous research with *P. cinereus* suggested that both sexes can remember encounters with unique individuals (e.g., see sections 7.10 and 7.12). This lead Jaeger et al. (2005) to test the hypotheses that single adult males and females of *P. cinereus* have short-term memory pertaining to events associated with a past territorial intrusion. They designed two experiments to test for displacement behavior in which an "event" induces a behavioral response that is directed (displaced) elsewhere. In these experiments, the event was a territorial intrusion by *Eurycea cirrigera* (experimental treatment) or a surrogate control followed by removal of these stimuli for 5 minutes and then the introduction of an unfamiliar conspecific of the opposite sex. They predicted that if males and/or females could remember the intrusion of the *E. cirrigera*, then those tested in the experimental treatment would be significantly more aggressive toward the subsequent conspecific intruder (i.e., displacement of territorial aggression) than the same individuals when tested in the control treatment. This response would indicate that the salamanders remembered the event of a previous territorial intrusion.

Adults of *P. cinereus* and 20 *E. cirrigera* were collected near MLBS and were tested in the laboratory at ULL where they were fed *D. virilis*. In treatment 1, focal males (experiment 1, $N = 28$) and females (experiment 2, $N = 27$) of *P. cinereus* were exposed to *E. cirrigera*. On day 1, Jaeger et al. (2005) placed each randomly chosen focal salamander into a Nunc bioassay chamber that contained a moist paper towel. They placed an empty screen cage ($10 \times 2.5 \times 1.5$ cm), 1 mm mesh with a solid bottom, on the left side of each chamber and fed each focal salamander eight fruit flies. On day 4, they fed the salamanders eight more flies and added more spring water if necessary. On day 5, they removed any unconsumed flies from each focal salamander's chamber and replaced the empty screen cage with an identical one containing a randomly assigned *E. cirrigera*. The cage allowed the focal salamander to see and detect chemical cues from this intruder but there was no direct contact. After 15 minutes, the focal salamanders had a 5-minute no-intruder period, followed by the introduction of a second intruder, *P. cinereus* (unfamiliar and opposite sex of the focal), into the focal salamanders' chamber for 15 minutes. Jaeger et al. recorded the aggressive behavior of both the focal and the second intruder salamander. *Plethodon cinereus* is known to be aggressive toward *E. cirrigera* (Jaeger et al., 1998, in section 6.3). In treatment 2, each focal *P. cinereus* was exposed to a surrogate control (rolled moist paper towel) placed into the cage on day 5 during the first intrusion. Data from males and females were considered separate experiments, and each focal salamander was tested twice in random order: once in each treatment. Conspecific intruders were used only once.

Focal males in the experimental group spent significantly more time in ATR (all trunk raised) toward conspecific female intruders than those in the control group. This result was not based on the responses of conspecific female intruders, because there was no significant difference in ATR by female intruders toward males that had been with *E. cirrigera* versus males that had been with surrogates first. Focal females did not significantly differ in ATR toward conspecific males after exposure to *E. cirrigera* or surrogate controls first. Male intruder behavior also did not differ significantly. These results suggest that at least males of *P. cinereus* have short-term memory of a territorial intrusion by *E. cirrigera*. Males (but not females) spent significantly more time in ATR toward conspecifics of the opposite sex after exposure to *E. cirrigera* than to controls. Jaeger et al. (2005) interpreted the males' behavior to be the displacement of territorial aggression from the original stimulus (*E. cirrigera*) to a subsequent stimulus (conspecific) after a 5-minute delay. This was the first test of short-term memory in *P. cinereus*.

9.4 THE IMPACT OF FAMILIARITY ON SALAMANDER BEHAVIOR

Memory involves the capacity to store, retain, and retrieve information from the past and can be studied by looking for a change in behavior due to experience (Dukas, 1998). Thus if *P. cinereus* changes its aggressive behavior toward previously encountered (familiar) individuals compared to unfamiliar individuals, then it must be able to remember those previously encountered individuals.

Jaeger and Peterson (2002) conducted a laboratory experiment to determine the influence of familiarity on the aggressive behavior of female *P. cinereus*. They tested three alternative hypotheses:

H_1: Females that are familiar and unfamiliar with each other are equally aggressive (the null), because females are in competition with each other for a limited food supply.

H_2: Females that are familiar with each other are less aggressive than are unfamiliar females.

H_3: Females that are familiar with each other are more aggressive than are unfamiliar females.

Single adult females ($N = 125$) were collected near MLBS and tested at ULL. The females were partitioned into groups of five (A–E) so that the future territorial resident (A: $N = 25$) was tested twice, in random order and 7 days apart, once with a familiar intruder (B) and once with an unfamiliar intruder (D). Females C and E were used to balance the experimental design. On day 1 of the familiar intruder trial, Jaeger and Peterson (2002) placed females A and B into a Nunc bioassay chamber and fed the females 20 *D. virilis*. On day 4, they fed the salamanders another 20 flies. On day 6, they transferred female B to a clean chamber while A stayed in the original chamber. On day 7, A and B were reunited in A's chamber. Both salamanders were placed under separate habituation cups for 15 minutes, and then the authors recorded the behavior of the salamanders for 15 minutes.

On day 1 of the unfamiliar intruder trial, Jaeger and Peterson (2002) placed females A and C into one chamber and females D and E into another chamber. The salamanders were fed 20 flies on days 1 and 4. On day 6, female C was removed from females A's chamber, while female E was removed from female D's chamber and then D was transferred to a new chamber. On day 7, female D was transferred into female A's chamber, both females A and D were placed under habituation cups for 15 minutes, and afterward the authors recorded the behavior of both salamanders for 15 minutes. There were 7 days between tests.

Jaeger and Peterson (2002) found that resident females spent significantly more time in ATR and nose taps (NT) in the presence of unfamiliar female intruders than in the presence of familiar female intruders. There was no significant difference in TOUCH. Unfamiliar female intruders spent significantly more time in ATR toward residents than did familiar intruders. There was no significant difference in NT or TOUCH between the two groups of intruders. Therefore, the data supported the second hypothesis that familiar females of *P. cinereus* are significantly less aggressive toward each other than are unfamiliar females, at least at MLBS. Familiarity may be a major factor in reducing aggressive contests among individuals. Thus female salamanders can recognize, remember, and distinguish the difference between familiar and unfamiliar individuals, and they can change their levels of aggression accordingly.

Joseph et al. (2005) also studied the role of familiarity on the behavior of *P. cinereus*. They examined how males and females develop an association with each other after they first meet. More specifically, they investigated if living together for 9 days is sufficient time to establish a socially monogamous relationship (see section 7.10). They asked if increasing familiarity between a male and a female leads to reduced aggression and increased touching, both of which were reported for socially monogamous pairs of *P. cinereus* in section 7.10. Joseph et al. hypothesized that familiarity contributes to the formation of socially monogamous pair associations.

Joseph et al. (2005) collected single adult males and females near MLBS and tested them in the laboratory at ULL. To start, they randomly selected males ($N = 27$) for experiment 1 and females ($N = 29$) for experiment 2 and placed them into Nunc bioassay chambers lined with moist paper towels. The focal salamanders were allowed 4 days to mark territories. After 4 days (on day 1 of the experiment), the authors introduced a randomly chosen salamander of the opposite sex into the chamber and placed this first intruder under a habituation cup for 15 minutes. After 15 minutes, the habituation cup was removed, and they recorded the behavior of the residents for 15 minutes. The resident and intruder were together until day 9. On day 9, the first intruder was placed under a habituation cup for 15 minutes, and afterward the authors recorded the behavior of the resident toward the first (now-familiar) intruder for 15 minutes. On day 10, they moved the first intruder to a clean Petri dish. Then they introduced a new randomly chosen second intruder of the opposite sex under a habituation cup for 15 minutes. Afterward they recorded the behavior of the resident for 15 minutes and then removed the second intruder. The residents were alone until day 12. On day 12, they reintroduced the first (now-familiar) intruder under a habituation cup for 15 minutes and then recorded the residents' behavior.

Both resident males and resident females were significantly more threatening toward the first opposite-sex intruder on day 1 than they were toward him or her on day 9; threat behavior significantly increased on day 10 toward the unfamiliar second intruder, then significantly decreased on day 12 toward the returning familiar first intruder. There was no significant difference in TOUCH across the days for both males and females. Therefore, the data supported the first hypothesis that male and female residents reduce their threat behavior toward an intruding member of the opposite sex over 9 days. This suggested that an increase in familiarity leads to a decrease in aggression by *P. cinereus*, at least near MLBS. Residents distinguished between familiar and unfamiliar intruders of the opposite sex, because residents increased aggression toward second intruders then decreased aggression toward returning first intruders. Also, Joseph et al. (2005) inferred that both males and females are able to retain the memory of the familiar first intruder over a 2-day separation even after encountering a different intruder (a "distractor," as in section 9.5) during those two days. They concluded that short-term familiarity does lead to reduced aggression between salamanders, but familiarity alone is not sufficient to explain the development of social monogamy in *P. cinereus*, as found in section 7.15. These results led to two new testable hypotheses:

H_1: Social monogamy in *P. cinereus* develops after long-term associations (>9 days) between opposite-sex individuals.

H_2: It requires a "choice" of associations by those individuals (as in section 7.15).

9.5 INDIVIDUAL RECOGNITION MEMORY

The results from the experiments on dear enemy recognition (Jaeger, 1981b, in section 3.5), familiarity (Jaeger & Peterson, 2002; Joseph et al., 2005, in section 9.4), and social monogamy (Gillette, Jaeger, et al., 2000; Jaeger, Gillette, et al., 2002; Prosen et al., 2004) led to a series of experiments on individual recognition memory.

9.5.1 The formation of individual recognition memory

Kohn et al. (2013) used different exposure durations (time that two individuals initially interacted) and separation intervals (time between interactions) to answer the following questions: How long do males have to interact before they become familiar with each other, and at what length of separation time do males behave toward familiar individuals as they do toward unfamiliar individuals (either because of "forgetting" or the need to reassess potential competitors)? They conducted three experiments to test how exposure duration impacts memory of individuals. Using experiments 3 and 4, they tested the consequences of two separation intervals. Kohn et al. predicted that longer exposure durations during first encounters would lead to reduced aggression toward familiar individuals during second encounters. Experiment 1, a 15-minute exposure duration, simulated an intruder merely passing through a focal salamander's territory. Experiment 2, an 8-hour exposure duration, simulated an individual that had recently established a territory nearby. Experiment 3, a 5-day exposure duration, simulated an established territorial neighbor. They also predicted that male salamanders are able to remember familiar individuals after a 5-day separation interval (experiment 3) based on the results from Kohn and Jaeger (2009). However, they had no predictions for whether salamanders could remember a familiar individual after a longer 15-day separation interval (experiment 4). They used aggressive behavior (ATR) as a way to assess recognition memory because *P. cinereus* tends to be less aggressive toward familiar intruders than toward unfamiliar intruders (Jaeger & Peterson, 2002; Joseph et al., 2005, in section 9.4).

All single adult male salamanders were collected near MLBS and tested in the laboratory at ULL. In experiment 1 (15-minute exposure and 5-day separation interval), Kohn et al. (2013) divided the salamanders into 26 groups of five. Within each group, they randomly selected one male as the focal salamander, and the others were randomly chosen as intruders. Each focal salamander went through three treatments: (1) familiar intruder trials, (2) unfamiliar intruder trials, and (3) surrogate control trials in random order, every third week until each focal salamander completed each treatment.

On day 1 of the familiar intruder trial, Kohn et al. (2013) placed the focal male ($N = 26$) into a Nunc bioassay chamber containing a moist paper towel. On day 5, they transferred male intruder A into the focal salamander's chamber for 15 minutes and allowed the salamanders to interact. After 15 minutes, they removed intruder A. On day 10, they reintroduced intruder A into the focal male's chamber and recorded the behavior of the focal male for 15 minutes. The unfamiliar intruder trials had similar procedures as the familiar intruder trials except, on day 5, intruder B was introduced into the focal male's chamber for 15 minutes, and on day 10, intruder was introduced for 15 minutes; researchers then recorded the behavior of the focal salamander. In surrogate control trials, intruder D was introduced into the focal male's chamber for 15 minutes on day 5. Then, on day 10, they introduced a surrogate control (moist rolled paper towel) into the focal male's chamber and recorded his behavior. All intruders were used only once.

In experiment 2 (8-hour exposure and 5-day separation interval), Kohn et al. (2013) randomly assigned 40 males to four groups: focal salamanders, future familiar intruders, future unfamiliar intruders, and the partner for the unfamiliar intruder (unfamiliar pair). Using a counterbalanced design, researchers tested each focal male in two treatments: familiar and unfamiliar intruder trials with one week between tests. On day 1 of the familiar intruder trials, they placed the focal male into a chamber containing a moist paper towel. On day 5, they introduced the familiar intruder into the focal male's chamber for 8 hours, after which they removed the familiar intruder and changed the substrate (to remove any remaining pheromones). On day 9, they changed the substrate again. On day 10, they reintroduced the familiar intruder into the focal male's chamber and recorded the focal male's behavior for 15 minutes. The unfamiliar intruder trials had the same procedures except on day 10, when an unfamiliar intruder was introduced into the focal male's chamber for 15 minutes. Also, unfamiliar intruders had been given the same social experience as familiar intruders, as they interacted with an unfamiliar pair for 8 hours on day 5.

In experiment 3 (5-day exposure and 5-day separation interval), Kohn et al. (2013) randomly assigned 40 males to four groups: focal salamanders, future familiar intruders, future unfamiliar intruders, and unfamiliar pairs. Using a counterbalanced design, they tested each focal male in two treatments: familiar and unfamiliar intruder trials with one week between treatments. On day 1 of the familiar intruder trials, they placed each focal male and its designated familiar intruder into a chamber containing a moist paper towel. On day 5, they removed the familiar intruder and placed him into a clean new chamber and also changed the substrate in the focal male's chamber. On day 9, they changed the substrate in the focal male's chamber again. On day 10, they reintroduced the familiar intruder into the focal male's chamber and recorded the behavior of the focal male for 15 minutes. The unfamiliar intruder trials had the same procedure except that on day 10 an unfamiliar intruder was introduced to the focal male. The unfamiliar intruders interacted with an unfamiliar pair for the first 5 days, and then they were introduced to the focal male on day 10.

In experiment 4 (5-day exposure and 15-day separation interval), Kohn et al. (2013) randomly assigned 40 males to the following four groups: focal salamanders, future familiar intruders, future unfamiliar intruders, and unfamiliar pairs. Using a counterbalanced design, they tested each focal individual in two treatments: familiar and unfamiliar intruder trials with one week between treatments. For the familiar intruder trial, days 1 through 5 were the same as in experiment 3. On day 19, the substrate was changed. On day 20, researchers reintroduced the familiar intruder into the focal male's chamber and recorded the behavior of the focal male. The unfamiliar intruder trial was the same as the familiar intruder trial except on day 20 an unfamiliar intruder was introduced into the focal male's chamber. Again, unfamiliar intruders had the same social experience as focal salamanders during days 1 through 5.

Thus Kohn et al. (2013) used different exposure durations and separation intervals to examine how long it takes for males to become familiar with each other and what length of separation is needed for familiar individuals to be responded to as if they are unfamiliar individuals. In experiment 1, focal males spent significantly more time in ATR toward both familiar and unfamiliar intruders than toward surrogate controls. There was no significant difference in ATR toward familiar intruders compared to unfamiliar intruders, nor were there significant treatment order effects for ATR in any of the four experiments. In experiments 2 and 3, focal males spent significantly more time threatening unfamiliar intruders than familiar intruders ($p = 0.002$ for experiment 2, $p < 0.0001$ for experiment 3). In experiment 4, there was no significant difference in ATR ($p = 0.144$) or number of bites toward ($p = 0.101$) familiar intruders compared to unfamiliar intruders. Therefore male salamanders can remember familiar conspecifics (neighbors) when the exposure duration is 8 hours or longer and the separation interval is short (5 days). This recognition memory is important for the salamanders at MLBS because the population is dense and there are many floater individuals that do not hold territories (Gillette, 2003; Mathis, 1991b). Thus Kohn et al. suggested that natural selection has acted on *P. cinereus* to discriminate between long-term territorial intruders/neighbors from short-term transient individuals. However, a lack of differences in behavior toward familiar and unfamiliar intruders (in experiments 1 and 4) does not directly indicate a lack of recognition memory. Memory is a neural representation of a past experience that can only be assessed by a change in behavior indicating that the animal has remembered that experience (Kawecki, 2010). Therefore, long separation intervals (15 days) may either lead to a loss of memory of previously familiar individuals or, alternatively, to aggressive reassessment of those individuals.

9.5.2 Sensory modalities used during recognition

Plethodon cinereus can remember familiar individuals, but what cues are used to remember those individuals? Kohn and Jaeger (2009) conducted two experiments with males to test whether territorial *P. cinereus* use only chemical or only visual cues to remember familiar conspecifics. They hypothesized that *P. cinereus* could recognize and remember individuals based on (1) just chemical cues or (2) just visual cues. Hypothesis 1 was based on Jaeger (1981b, in section 3.5), who found that *P. cinereus* can use scent marks for dear enemy recognition without visual or physical contact. Kohn and Jaeger posited hypothesis 2 because *P. cinereus* uses visual displays while defending territories (Jaeger, 1984) and uses vision during feeding (David & Jaeger, 1981; Jaeger, Barnard, et al., 1982).

In experiment 1, on chemical cue memory, Kohn and Jaeger (2009) collected single males near MLBS and tested them in the laboratory at ULL. They randomly assigned 40 males to each of the four groups: focal salamanders, future familiar intruders, future unfamiliar intruders, and unfamiliar pairs. The experiment was counterbalanced, and each focal male was tested in two treatments: familiar intruder and unfamiliar intruder trials with one week between treatments. On day 1 of the familiar intruder trails, researchers placed each focal male with its designated intruder into a Nunc bioassay chamber containing a moist paper towel so that the two salamanders could become familiar with each other over 5 days. On day 5, they removed the familiar intruder and placed him in a clean Petri dish containing one moist coffee filter and one moist 9 cm filter paper. They also changed the substrate in the focal salamander's chamber. On day 10, just before testing, they rolled the 9 cm filter paper from either the familiar or the unfamiliar intruder's Petri dish and placed it into the center of the focal male's chamber for 15 minutes while recording the focal male's behavior. So that unfamiliar intruders could have the same social experience as familiar intruders, unfamiliar intruders interacted with an unfamiliar pair during days 1 through 5, then were separated from their unfamiliar pair and placed into a Petri dish with filter papers on day 5; their chemical cues were introduced to the focal salamander on day 10.

In experiment 2, on visual cue memory, Kohn and Jaeger (2009) randomly assigned 36 males to the following groups: focal males, future familiar intruders, future unfamiliar intruders, and unfamiliar pairs. Each focal male was tested in three treatments in random order: familiar intruder trials, unfamiliar intruder trials, and surrogate control trials with one week between treatments. The researchers randomly assigned the order of presentation. On day 1 of the familiar intruder trials, they placed the focal individual and its designated familiar intruder into a chamber containing a moist paper towel. On

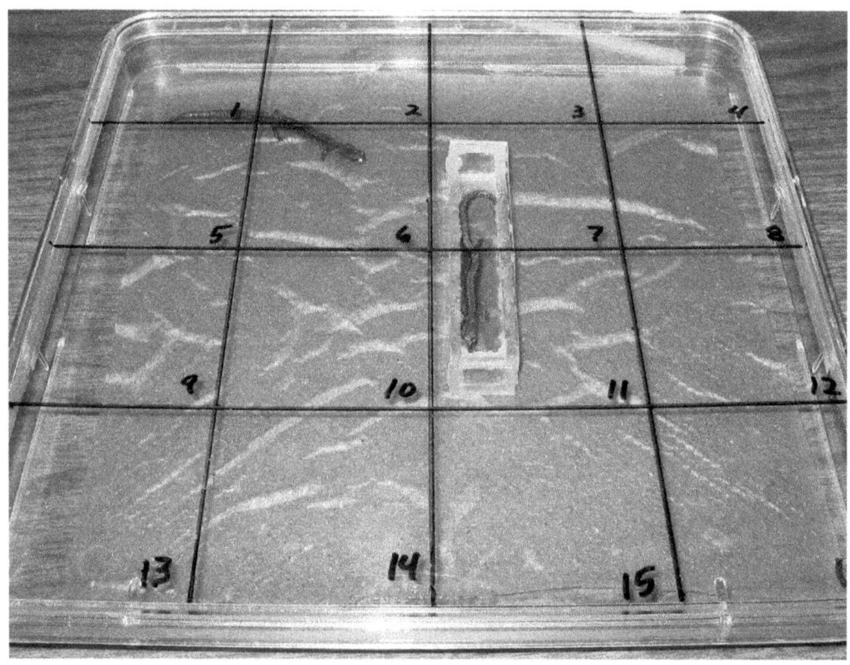

Figure 9.2 The visual chamber used in Kohn and Jaeger (2009) that allowed the focal salamander to see the intruder but not chemically detect him.

day 5, they removed the familiar intruder and placed him in a clean chamber containing a moist paper towel. They also changed the substrate in the focal male's chamber. On day 10, they placed either the familiar intruder, the unfamiliar intruder, or a surrogate control into a clear glass chamber (a "visual chamber" 83 × 14 × 9 mm), which was sealed with Vaseline so that it was airtight (Fig. 9.2). They then placed the visual chamber into the center of the focal male's chamber and recorded his behavior for 15 minutes. Unfamiliar intruders interacted with an unfamiliar pair during days 1 through 5, were separated on day 5, and were introduced (in the visual chamber) to the focal salamander on day 10.

In the chemical cue experiment, focal males spent significantly more time in ATR toward the rolled paper towel containing the chemical cues of unfamiliar intruders than those of familiar intruders ($p < 0.0001$). Focal males also NT significantly more often toward the rolled filter paper during unfamiliar intruder trials than familiar intruder trials ($p = 0.03$). There were no treatment order effects. In the visual cue experiment, focal males differed significantly across the three treatments ($p < 0.01$). Focal males spent significantly more time in ATR toward unfamiliar intruders than toward either familiar intruders or surrogate controls. There was no significant difference in NT to the substrate ($p > 0.30$), no treatment order effects, and no intruder salamanders displayed ATR during the 15 minutes of observations.

This research implies that male salamanders can use only chemical cues or only visual cues for individual recognition memory. Kohn and Jaeger (2009) were the first to examine whether salamanders use both odor and visual cues for recognizing individuals, and they suggested that chemical cues are more important than visual cues. Individual recognition memory is probably important for territorial salamanders at MLBS because of (1) the potential for frequent social encounters with other individuals and (2) the complex forest floor environment in which they live. It is also beneficial for recognizing long-term territorial neighbors and treating them differently than transient individuals moving through an environment (Kohn et al., 2013; Palphramand & White, 2007).

9.5.3 Do distractors inhibit individual recognition memory?

These recognition memory experiments dealt with a salamander remembering a single individual with no other social interactions during the separation interval (except for Joseph et al., 2005, in section 9.4), but *P. cinereus* on the forest floor interacts with many other salamanders at MLBS (see section 7.19). Therefore, Kohn (2008) investigated whether encounters with unfamiliar conspecifics ("distractors") during the separation interval impacted the memory of the original familiar individual in male *P. cinereus*. In experiment 1, she compared the recognition memory of familiar conspecifics after the focal salamander interacted with zero, one, or three unfamiliar males during the separation interval. In experiment 2, she compared levels of aggression (as an indicator of memory) toward familiar and unfamiliar intruders after focal males interacted with one or three other male conspecifics.

Kohn (2008) collected 295 single adult males near MLBS and tested them in the laboratory at ULL. In experiment 1, she randomly assigned 42 males as focal salamanders; all other salamanders were randomly assigned as familiar intruders or unfamiliar intruders. Using a counterbalanced design, she tested each focal salamander in three treatments: (1) zero intruding unfamiliar salamanders, (2) one intruding unfamiliar salamander, and (3) three intruding unfamiliar salamanders with one week between treatments. On day 1 of treatment 1, she placed the focal and designated familiar intruder into a chamber containing a moist paper towel. On day 5, she removed the familiar intruder and placed him in his own chamber containing a moist paper towel and changed the substrate in the focal salamander's chamber. On days 6 through 8, she changed the substrate in the focal salamander's chamber. On day 10, she reintroduced the familiar intruder into the focal male's chamber and recorded the behavior of the focal salamander for 15 minutes.

For treatment 2 (one intruding unfamiliar salamander), the procedures were the same as treatment 1 except on day 8 Kohn (2008) placed a never-before-seen, unfamiliar intruder into the focal male's chamber for 8 hours. After 8 hours the unfamiliar intruder was removed and Kohn changed the substrate in the focal male's chamber. On day 10, she reintroduced the familiar intruder into the focal salamander's chamber and recorded the behavior of the focal male for 15 minutes. For treatment 3 (three intruding unfamiliar salamanders), the procedures were the same as previously on days 1 through 5. On days 6, 7, and 8, Kohn placed a different unfamiliar intruder, one each day, into the focal male's chamber and left them there for 8 hours. After 8 hours, the unfamiliar intruders were removed and the substrate in the focal male's chamber was changed. On day 10, she reintroduced the familiar intruder into the focal male's chamber and recorded the behavior of the focal male for 15 minutes. She found no significant difference in ATR toward the familiar intruders across the three treatments. She also found no treatment order effect. These nonsignificant results led to experiment 2, in order to determine if males can remember a familiar intruder after interacting with one or three other salamanders during the separation interval and to determine if recognition memory changes if the salamanders interact with more individuals during the separation interval.

In experiment 2, Kohn (2008) collected 316 single adult males near MLBS and tested them in the laboratory at ULL. She randomly assigned males to treatment groups using the following criteria: (1) males could only be used as familiar intruders once; (2) once used as a familiar intruder the males could be used as an unfamiliar intruder, unfamiliar pair, or an intruding (distractor) salamander for another focal individual; and (3) salamanders were allowed at least 6 days with no interactions before being used again in the experiment. Treatment groups included focal salamanders, familiar intruders, unfamiliar intruders, unfamiliar pairs, and intruding distractor salamanders, with 36 males assigned as focal males. She tested each focal salamander in four treatments: (1) one intruding distractor with the original familiar intruder, (2) one intruding distractor with an unfamiliar intruder, (3) three intruding distractors with the original familiar intruder, and (4) three intruding distractors with an unfamiliar intruder with one week between treatments and the order of treatments randomly assigned.

On day 1 of treatment 1, Kohn (2008) placed both the focal salamander and its designated familiar intruder into a chamber containing a moist paper towel. On day 5, she removed the familiar intruder and placed him in a clean new chamber with a moist paper towel and changed the substrate in the focal male's chamber. On days 6 and 7, she changed the substrate in the focal male's chamber again. On day 8, she placed an intruding distractor salamander into the focal salamander's chamber for 8 hours. After 8 hours, the intruding distractor salamander was removed and the substrate in the focal male's chamber was replaced. On day 10, she reintroduced the original familiar intruder into the focal male's chamber and recorded the behavior of the focal male for 15 minutes. The procedures were the same for treatment 2, except on day 10 Kohn placed an unfamiliar intruder into the focal male's chamber and recorded the behavior of the focal salamander. In order for the unfamiliar intruder to have a similar social experience as the focal salamander, each unfamiliar intruder interacted with an unfamiliar pair on days 1 through 5, was separated on day 5, and then introduced to the focal male on day 10.

The procedures for treatment 3 were the same on days 1 through 5. On days 6, 7, and 8, Kohn (2008) placed an intruding distractor salamander (a different one each day) into the focal male's chamber for 8 hours. After 8 hours, each distractor salamander was removed and the substrate in the focal male's chamber was changed. On day 10, she reintroduced the familiar intruder into the focal male's chamber and recorded the behavior of the focal male for 15 minutes. The procedures for treatment 4 were the same as treatment 3, except on day 10 Kohn placed an unfamiliar intruder into the focal male's chamber and recorded the behavior of the focal salamander for 15 minutes. Unfamiliar intruders had a similar social experience as familiar intruders during days 1 through 5 as they interacted with an unfamiliar pair were separated on day 5, and then were introduced to the focal salamander on day 10.

Kohn (2008) examined whether males could remember a familiar intruder after interacting with one or three other salamanders (distractors) during the separation interval. She found that focal males spent significantly more time threatening unfamiliar intruders when there was one ($p = 0.001$) and three ($p = 0.0001$) intruding distractor salamanders than threatening familiar intruders. Also, focal males spent significantly more time threatening returning familiar intruders after interacting with one intervening distractor salamander than threatening returning familiar intruders after interacting with three intervening intruding distractor salamanders ($p = 0.02$). There was no treatment order effect. Therefore male salamanders can remember familiar individuals after encountering one or three other intruding salamanders during a 5-day separation period. This is not surprising considering the potential for frequent social encounters with other individuals at our study site near MLBS (see section 7.19). What was surprising was that focal males were significantly less aggressive toward familiar conspecifics after interacting with three intruding distractor salamander than after interacting with just one intruding distractor salamander during the separation period, because memory theory predicts that "forgetting" is more likely to occur as the number of distractors increases due to a greater probability of interference with the memory (Koster et al., 2002).

9.5.4 Overview

In summary, we found that learning, memory, and decision-making (e.g., levels of aggression) play important roles in the social behavior of *P. cinereus*. Their territorial lifestyle has resulted in many adaptations, such as a decrease in aggression among territorial adults due to familiarity and the ability to remember familiar individuals, after at least an 8-hour interaction for at least 5 days. *Plethodon cinereus* can also remember a familiar individual after interacting with up to three other conspecifics and uses both chemical and visual cues for recognition memory. Moreover, males, but not females, show displacement behavior of territorial aggression from an original stimulus to a subsequent stimulus after a 5-minute delay. Similar to human infants, *P. cinereus* can discriminate between 1 versus 2 and 2 versus 3 food items but they cannot discriminate numbers higher than 3. We also show that their foraging behavior has both a heritable (or at least maternal) component and a learning component for neonates, juveniles, and adult stages. Thus learning, memory, and decision-making are important factors in the cognitive ecology of this species.

9.6 SELECTED RECENT RESEARCH BY OTHERS: COGNITIVE ECOLOGY

The ground-breaking research by Uller et al. (2003) on numerical discrimination in *P. cinereus* was the first to show that amphibians (sometimes considered lower vertebrates) can discriminate quantities of up to 2 versus 3. Using larger sets of numbers, Krusche et al. (2010) found that *P. cinereus* could also discriminate between 8 versus 16 prey items, which falls into the 1:2 ratio of numerosity often used by other species. Because the authors tested foraging preference using both live and computer-animated food items, they were able to determine that the salamanders were assessing quantities on the basis of movement-related cues. However, variables that covary with number (quantitative variables) such as volume and surface area that can also affect discrimination were not controlled for in these studies. Stancher et al. (2015), using similar free-choice experiments with prey items, found that frogs (*Bombina orientalis*) can discriminate quantities up to 2 versus 3 and larger numerousness of 3 versus 6 and 4 versus 8, which also falls into the 1:2 ratio of numerosity. They also controlled for quantitative variables and found that, for small numerousness (i.e., 1 vs. 2), neither the number of prey items nor surface areas of the prey items were dominant factors associated with quantity-discrimination. They proposed that when faced with larger numerousness, using the total surface area of a group may represent the best strategy for selection between prey or other relevant resources. Guppies (*Poecilia reticulata)* also discriminate numerosities in a 1:2 ratio during a food choice test (Lucon-Xiccato et al., 2015). Researchers found that guppies attend to cumulative surface area of food items rather than number of items to select the larger quantity. They also found that guppies prioritize larger items over cumulative surface area of food. Lucon-Xiccato et al., hypothesized that because guppies forage in social groups (shoals), this strategy may maximize foraging success. Performing a similar study with salamanders would provide a useful test of this hypothesis. Further, testing larval salamanders may open new doors, because in pond-dwelling species we expect foraging competition to play a greater role than in terrestrial species.

There are other interesting directions to take with salamanders. One is a comparative approach across species to learn if they share the same numerical abilities. Agrillo et al. (2012) did this with five teleost species of fish. They trained fish, using a food reward, to discriminate between two sets of geometrical figures that differed in numerosity. Once the fish reached a learning criterion ratio of 1:2, they were tested in nonreinforced trials for their ability to generalize more difficult ratios (0.67) and different set sizes. The authors found that five species of fish generalized numerosities with a 0.67 ratio but only with smaller-size sets. Many studies in cognition, including numerical discrimination, include a training procedure before the test, which is different from the spontaneous-choice test design in Uller et al. (2003). Agrillo and Bisazza (2014) reviewed the pros and cons of the two methodologies and argued that a combined approach of training first and then exploring spontaneous discrimination may provide greater insight into numerical abilities of a given species.

10

Coda

Synthesis and social behaviors by P. cinereus

The researches summarized in the book cover 114 publications, 6 dissertations, and 3 theses just concerning *P. cinereus* and just by our research group. The early publications (1970–1985) established a "hard core" theory (*sensu* Lakatos, 1970) of the basic ecology and social behaviors of *P. cinereus*, as reviewed by Jaeger and Forester (1993) and Mathis et al. (1995). From this hard core approach sprang subsequent research projects, many led by graduate students, that examined such diverse areas as geographic variation and complex social behaviors in *P. cinereus*. In addition, another 112 manuscripts came forth from rebellious students who were "bored with *P. cinereus*." These dealt with ecology and/or social behaviors of species ranging from gastropods to pronghorn antelopes, including many publications on other species of salamanders. These projects resulted in fruitful exchanges of methodologies between "pro-" and "anti-" *P. cinereus* students.

10.1 BEHAVIORAL VARIATION WITHIN A POPULATION

We now explore six variables that may have led to behavioral variation among individuals of *P. cinereus* in their territorial and other social behaviors within a given experiment. Such variation caused statistical nightmares (see section 1.4) but were important because they suggested that different individuals may have different (genetic and/or learned) behavioral tactics in responding to a common set of external stimuli. For example, during the foraging experiments in chapter 4, we observed that some individuals consistently used ambush tactics while preying on *Drosophila* despite different densities of those flies. Others slowly stalked the flies, while still others ran after prey or even ran and leapt into the air to capture a fly, but most salamanders varied their tactics depending on the densities of the flies.

We have already shown in section 3.10 that territorial behavior of *P. cinereus* varies geographically among populations and seasonally within a population. However, our experiments could not control for unknown traits among individuals within a single population such as at Mountain Lake Biological Station (MLBS). We suggest a few interindividual differences that probably influence the behaviors of *P. cinereus*.

10.1.1 Age

We usually experimentally tested adults, and they were a minimum of 2 years old and a maximum of, we posit, 20 to 25 years old. Social and competitive behaviors may change with the physical condition of age and with prior learning experiences during aging.

10.1.2 Tail autotomy

In section 5.8, we showed that tail autotomy drastically alters the behaviors of territorial residents and their intruders. We controlled for this by collecting only tail-intact salamanders for most experiments, but we could not control for any lingering behavioral changes brought on by prior, historical autotomies (see section 7.19). *Plethodon cinereus* regrows the tail after autotomization, but tail loss to predators or during territorial conflicts could have long-term effects on future behaviors, especially in combats between conspecifics.

10.1.3 Polymorphism

At MLBS, *P. cinereus* is highly polymorphic in stripes, colors, and patterns. Some individuals possess a red stripe from head to tip of tail while others have broken stripes of various colors (from red to tan) and a few have no stripe at all: the "leadback" morph. In section 9.5, Kohn and Jaeger (2009) found that males of this species can remember familiar male individuals by visual information alone (even in the absence of pheromonal signals) and vary their behaviors accordingly. We do not know if *P. cinereus* possesses color vision, although frogs do (e.g., Hailman & Jaeger, 1974; Jaeger & Hailman, 1971). However, salamanders should at least be able to distinguish among patterns of stripes. Anthony et al. (2008) found evidence in Ohio that *P. cinereus* tends toward assortative mating (or at least toward assortive associations) based on color polymorphism. Also, Reiter et al. (2014) reported that in Ohio striped *P. cinereus* as residents was more aggressive and less submissive than were unstriped (leadback) residents during territorial contests. For our experiments, we could not collect salamanders of just a single morph due to the vast variation in stripes, so these multimorphs may have contributed to the behavioral variation in our data.

10.1.4 Health

Although we have never seen ectoparasites on *P. cinereus*, we could not otherwise distinguish healthy from unhealthy salamanders collected in forests and used in the experiments, and this may explain some of the behavioral variation in the data. For example, in any given test, some individuals invariably spent the entire 15 to 30 minutes in the FLAT posture while others actively moved and interacted (e.g., all trunk raised [ATR]) with conspecifics during the entire observation. Therefore a healthy salamander may behave quite differently from an unhealthy conspecific.

10.1.5 Bold and shy

We observed individual consistencies in behavior of *P. cinereus* within ("behavioral type") and across different contexts ("behavioral syndromes": Sih et al., 2004). Some salamanders consistently would BITE (suggestive of "bold" behavior) and others would not (suggestive of "shy" behavior) during behavioral interactions, implying that behavioral types play a role in the variation in behavior observed during our research. Crane et al. (2012) found that Ozark zigzag salamanders, *Plethodon angusticlavius,* showed consistent bold and shy responses to predator cues. Specifically testing whether *P. cinereus* shows behavioral types during territorial interactions or revisiting our old datasets might provide further insight into whether *P. cinereus* truly shows behavioral syndromes, because our research frequently consisted of testing salamanders multiple times across contexts.

10.1.6 Experimental flaws

It is possible that the primary contributor to behavioral variances comes from stressing the salamanders before or during an experiment. Woodley and Lacy

(2010) found that plasma corticosterone increased and salamander activity decreased after being handled in a plastic bag, shaken for three or four times, and having their sex evaluated, in another plethodontid species (*Desmognathus ocoee*). However, when salamanders were transferred to testing chambers with minimal handling, they did not decrease activity. Bliley and Woodley (2012) further measured the effects of repeated handling on *D. ocoee* and found that after at least 9 days of daily handling, locomotory activity was reduced compared to a control treatment group, for both males and females. From our observations, *P. cinereus* when treated roughly, will spend most or all of an observational period in either FLAT or EDGE, yielding meaningless data.

We discovered three techniques that appear to reduce stress in *P. cinereus*. First, when transporting a salamander from one chamber to another, we allow it to walk willingly into a plastic tube in the first chamber and then walk out into the second chamber. This avoids all human touches to the salamander. Second, once in an experimental chamber (Fig. 3.4), the salamander should be allowed to habituate for at least 15 minutes before an observation begins. In a territorial experiment, for example, we place a small Petri dish over the resident and another dish over the intruder. This habituation period seems adequate for the salamanders to behave "normally" in terms of foraging, movements, and displays. Third, individuals of *P. cinereus* seem to be undaunted by humans staring into their chambers as long as the humans are peripheral to the chamber (i.e., not directly overhead), moving slowly, and with hand movements invisible to the salamander. We quickly learned that salamanders do not exist merely to test our hypotheses; instead they have their own needs to which behavioral ecologists must respond with understanding.

10.2 BEHAVIORAL OPTIONS DURING CONTESTS

Plethodon cinereus responds behaviorally to intra- and interspecific stimuli in a number of ways that depend on the context. Although variation among individuals tends to obscure our understanding of how these behavioral decisions are formulated, we have observed some general "rules" by which the salamanders, on average, operate. To illustrate these rules, we focus on an adult territorial resident, either male or female, confronted with an intruder entering its territory.

Option 1: Respond with high threat postures (ATR 3, 4, and 5; Fig. 3.1) before launching a biting attack. We call this an "honest threat signal," because it signals a forthcoming attack.

Option 2: Respond with ATR 3, 4, and 5 but fail to launch an attack. This is a "deceptive threat signal" or a "bluff."

Option 3: Respond with low threat postures (ATR 1 and 2) before launching a "sneak attack." This is rarely seen.

Option 4: Respond with ATR 1 and 2 but fail to attack. This is an "alert posture."

Option 5: Remain in the resting posture (front of trunk raised [FTR]) while looking toward (LT) the intruder, apparently while assessing the intruders intentions.

Option 6: Remain in FTR while not LT the intruder, or ignoring the intruder.

Option 7: Assume the FLAT submissive posture, which seems to inhibit an attack by the intruder.

Option 8: Flee from the intruder (move away [MA] or EDGE), as if abandoning its territory.

Option 9: Respond with a series of changing postures (e.g., ATR 1–5) and movements (e.g., MT, MA, EDGE) as if assessing the physical characteristics (RHP) and behavioral responses of the intruder.

Option 10: Enacted toward a predator, such as a snake: Remain FLAT and immobile, perhaps autotomize the tail, and then quickly flee.

Our impressions of these 10 options came from several thousands of hours of observing behavioral interactions, in chambers, of *P. cinereus* over four decades, as reviewed in chapters 2 through 8.

10.3 HOW SALAMANDERS CHOOSE AMONG OPTIONS

The option, or options, "chosen" by a salamander depend on the information that a resident (and its intruder) perceives about the intruder (and the resident) and about the quality of the territory. One can view this interaction between resident and intruder as a "game of assessments" (*sensu* Maynard Smith, 1982), by changing postures (ATR 1–5, FTR, FLAT) and movements (LT, LA, MT, MA). Apparently *P. cinereus* can perceive a vast number of variables, manipulate this diverse information in its neural network, and then "decide" on one or more behavioral responses. In the experiments discussed in chapters 2 through 8, these variables (as hypotheses) were tested only one or a few at a time, so we still do not understand how behavioral decisions are formulated based on a unified set of variables.

Table 10.1 provides our generalized understanding of how a salamander responds to just one variable at a time. Table 10.1 focuses on an adult, male, territorial resident confronted by an intruder. We choose a male here because, at any given time, about half of the females in a population would be gravid (yolking ova) while the other half would not be gravid (brooding or recovering from brooding), which influences their behavioral responses (see section 7.19). Table 10.1 provides only a few examples of decision-making by a resident *P. cinereus*, and it does not account for the variable postures and movements of the intruder, which will influence the resident's responses.

Table 10.1 RESPONSES BY AN ADULT, MALE, TERRITORIAL RESIDENT TO VARIOUS TYPES OF INTRUDERS AND TERRITORIAL QUALITY.

The intruder is	Resident's response is	Related section
I. Conspecific		
1. Same sex	High aggression	7.11
Opposite sex	Low aggression	7.11
2. Familiar	Low aggression	7.9
Unfamiliar	High aggression	7.9
3. Socially monogamous partner	Low aggression, resting	7.10
Socially polyandrous partner	High aggression	7.10
• Previously cooperative	Low aggression, resting	7.14
• Previously uncooperative	High aggression	7.14
4. Larger than resident	Low aggression	3.8, 3.9, 5.7, 6.2, 7.1
Same size	Lower aggression	3.8, 3.9, 5.7, 6.2, 7.1
Smaller than resident	Higher aggression	3.8, 3.9, 5.7, 6.2, 7.1
5. Tail intact	Low aggression	3.8, 3.10, 7.19
Tail autotomized	High aggression	3.8, 3.10, 7.19
II. Congener		
6. Juvenile *P. glutinosus*	Low aggression	6.2
Adult *P. glutinosus*	Resting	6.3
7. *P. shenandoah*	High aggression	6.1
8. *P. hubrichti*	High aggression	6.7
9. *P. hoffmani*	High aggression	6.6
III. Confamilial		
10. *Desmognathus fuscus*	Fleeing	6.3, 6.4
11. *D. ochrophaeus*	Fleeing?	6.5
12. *Eurycea cirrigera*	High aggression	6.3, 6.4
IV. Predatory snake		
13. *Diadophis punctatus*	Tail autotomy, fleeing	8
Resident's Territorial Quality		
14. Established for		
• <5 days	Low aggression	3.8
• ≥5 days	High aggression	3.8
15. Prey-rich	High aggression	3.8
Prey-poor	Low aggression	3.8
16. Excessively		
• Hot (>18°C)	Fleeing underground	3.2, 3.7
• Cold (<3°C)	Fleeing underground	3.2
• Dry	Go to cover object	3.8, 7.19

NOTE. High aggression = ATR 3–5 and MT; low aggression = ATR 1 and 2 and LT; resting = FTR; submission = FLAT and LA; fleeing = EDGE and MA.

10.4 DEFINITIONS OF SOCIAL, MATING, AND GENETIC MONOGAMY

We have casually used, without defining our usage, the terms *social, mating,* and *genetic monogamy* and their alternatives: *polygyny, polyandry,* and *polygamy.* We do not yet understand how these terms apply within populations of *P. cinereus*; while social monogamy was inferred from laboratory experiments (sections 7.10–7.12, 7.14, and 7.15) and observations in the forest (sections 7.1 and 7.19), DNA analyses from MLBS found rampant genetic polyandry in the forest (section 7.16). Now we suggest the many ways that female–male interactions may occur within the population at MLBS.

We define, for *P. cinereus, mutual social monogamy* as sharing a territory or overlapping territories while displaying favorable behavior (e.g., low or no threat displays) *uniquely* between two conspecific, adult, opposite-sex, nonkin individuals, which we term *partners* (*sensu* Gillette, Jaeger, et al., 2000, in section 7.10). This definition dismisses favorable behavior between familiar male–male and female–female individuals, by individuals of *P. cinereus* toward other species such as adults of *P. glutinosus*, and due to kin selection. We infer from section 7.19 that when social monogamy does occur, it lasts for only one reproductive cycle.

Mutual social monogamy is interesting because it increases the probability that a particular male and female will mate (i.e., sperm transfer). The alternatives to (1) mutual social monogamy are (2) male social monogamy/female social polyandry, (3) female social monogamy/male social polygyny, and (4) social polygamy.

Note that social relationships do not necessarily lead to matings in any of these four scenarios. The mating options are (5) mutual mating monogamy, (6) male mating monogamy/female mating polyandry, (7) female mating monogamy/male mating polygyny, and (8) mating polygamy.

However, mating does not necessarily lead to genes passed onto offspring. For example, a male may mate but be sterile (no viable sperm) or have a low sperm count due to previous matings; a female may mate but employ cryptic female choice (Eberhard, 1996) by manipulating sperm from one or more males in her spermatheca or herself be sterile (no viable ova). Therefore, the genetic outcomes of mating would be (9) mutual genetic monogamy (one father and one mother for a clutch of eggs), (10) male genetic monogamy/female genetic polyandry (several fathers for one clutch of eggs as in section 7.16), (11) female genetic monogamy/male genetic polygyny (male is father to all eggs in one clutch, but also fathers all or parts of other females' clutches), (12) genetic polygamy, and (13) genetic sterility (no viable eggs produced by a given male or female).

Which combination of social, mating, and genetic tactics occurs at MLBS probably depends on two major factors. First, the distribution of cover objects, and thus territories, would determine the options available to *P. cinereus*. When cover objects are closely spaced, territoriality appears to break down (section 7.2), perhaps leading to polygamy. However, when cover objects are very far apart, a male and female lucky enough to find a cover object may be forced into mutual social and mating monogamy (section 7.2). Where cover objects are close, but not too close, opportunities would occur for social monogamy (as in section 7.19) with extra-pair polygamy, polygyny, or polyandry (as in section 7.16).

Second, long-term associations between a male and a female in a forest probably enhance the probability of social, if not mating and genetic, monogamy, as suggested in section 7.15. A male and a female in adjacent territories may opt for an overlapping territory (section 3.7) or for a shared single territory (section 7.19), leading to some form of monogamy. All of these variables probably occur simultaneously at MLBS (or in any forest), varying among localities in the forest.

10.5 MEA MAXIMA CULPA

When this research program began in 1965, the senior author asked two apparently simple questions that he thought would yield simple answers.

1. What ecological and/or behavioral factors lead to the interspecific, competitive superiority of *P. cinereus* over some other same size *Plethodon* (particularly *P. shenandoah*)?
2. What social behaviors allow *P. cinereus* to maintain such large population densities in some forests (e.g., at Shenandoah National Park and MLBS)?

After 50 years of research, neither a complete, nor even satisfactory, answer has been provided to either question. The approximately 200 experiments and studies in the forests reviewed in this book have produced a jigsaw puzzle of *P. cinereus*, with only a few of the pieces fitting together. We hope that younger colleagues now studying *P. cinereus* will complete the jigsaw puzzle, or at least add new pieces to it: a philosophy of science favored by Popper (1959), Platt (1964), and Lakatos (1970).

REFERENCES

Acord, M. A., Anthony, C. D., & Hickerson, C.-A. M. (2013). Assortative mating in a polymorphic salamander. *Copeia*, 2013, 676–683.

Adams, D. C. (2007). Organization of *Plethodon* salamander communities: Guild-based community assembly. *Ecology*, 88, 1292–1299.

Adams, D. C., & Anthony, C. D. (1996). Using randomization techniques to analyse behavioural data. *Animal Behaviour*, 51, 733–738.

Agrillo, C., & Bisazza, A. (2014). Spontaneous versus trained numerical abilities. A comparison between the two main tools to study numerical competence in non-human animals. *Journal of Neuroscience Methods*, 234, 82–91.

Agrillo, C., Petrazzini, M. E. M., Tagliapietra, C., & Bisazza, A. (2012). Inter-specific differences in numerical abilities among teleost fish. *Frontiers in Psychology*, 3, 1–9.

Anthony, C. D., Hickerson, C.-A. M., & Venesky, M. D. (2007). Responses of juvenile terrestrial salamanders to introduced (*Lithobius forficatus*) and native centipedes (*Scolopocryptops sexspinosus*). *Journal of Zoology*, 271, 54–62.

Anthony, C. D., & Pfingsten, P. A. (2013). Eastern red-backed salamander *Plethodon cinereus* (Green 1818). In R. A. Pfingsten, J. G. Davis, T. O. Matson, G. Lipps, Jr., D. Wynn, & B. J. Armitage (Eds.), *Amphibians of Ohio* (pp. 335–360). Ohio Biological Survey Bulletin New Series 17. Columbus: Ohio Biological Survey.

Anthony, C. D., Venesky, M. D., & Hickerson, C.-A. M. (2008). Ecological separation in a polymorphic terrestrial salamander. *Journal of Animal Ecology*, 77, 646–653.

Arif, S., Adams, D. C., & Wicknick, J. A. (2007). Bioclimatic modelling, morphology, and behaviour reveal alternative mechanisms regulating the distributions of two parapatric salamander species. *Evolutionary Ecology Research*, 9, 843–854.

Aucoin, S., Jaeger, R. G., & Giambrone, S. (2005). Philosophy, value judgements, and declining amphibians. In M. Lannoo (Ed.), *Amphibian Declines: The Conservation Status of United States Species* (pp. 10–14). Berkeley: University of California Press.

Bliley, J. M., & Woodley, S. K. (2012). The effects of repeated handling and corticosterone treatment on behavior in an amphibian (Ocoee salamander: *Desmognathus ocoee*). *Physiology and Behavior*, 105, 1132–1139.

Bobka, M. S., Jaeger, R. G., & McNaught, D. C. (1981). Temperature dependent assimilation efficiencies of two species of terrestrial salamanders. *Copeia*, 1981, 417–421.

Brown, C. W. (1968). Additional observations on the function of the nasolabial grooves of plethodontid salamanders. *Copeia*, 1968, 728–731.

Brown, J. L., & Orians, G. H. (1970). Spacing patterns in mobile animals. *Annual Review of Ecology and Systematics*, 1, 239–262.

Buchanan, B. W., & Jaeger, R. G. (1995). Amphibians. In B. E. Rollin (Ed.), *The Experimental Animal in Biomedical Research*, Vol. 2 (pp. 31–48). Boca Raton, FL: CRC Press.

Burton, T. M., & Likens, G. E. (1975). Salamander populations and biomass in the Hubbard Brook Experimental Forest, New Hampshire. *Copeia*, 1975, 541–546.

Carpenter, D. W., Jung, R. E., & Sites, J. W. (2001). Conservation genetics of the endangered Shenandoah salamander (*Plethodon shenandoah*, Plethodontidae). *Animal Conservation*, 4, 111–119.

Chouinard, A. J. (2012). Rapid onset of mate quality assessment via chemical signals in a woodland salamander (*Plethodon cinereus*). *Behavioral Ecology and Sociobiology*, 66, 765–775.

Clutton-Brock, T. H., & Parker, G. A. (1995). Sexual coercion in animal societies. *Animal Behaviour*, 49, 1345–1365.

Courtene-Jones, W., & Briffa, M. (2014). Boldness and asymmetric contests: Role- and outcome-dependent effects of fighting in hermit crabs. *Behavioral Ecology*, 25, 1073–1082.

Crane, A. L., & Mathis, A. (2011). Landmark learning by the Ozark zigzag salamander *Plethodon angusticlavius*. *Current Zoology*, 57, 485–490.

Crane, A. L., McGrane, C. E., & Mathis, A. (2012). Behavioral and physiological responses of Ozark zigzag salamanders to stimuli from an invasive predator: The armadillo. *International Journal of Ecology*, 2012. doi:10.1155/2012/658437

Dalton, B., & Mathis, A. (2014). Identification of sex and parasitism via pheromones by the Ozark zigzag salamander. *Chemoecology*, 24, 189–199.

Dantzer, B. J., & Jaeger, R. G. (2007a). Detection of the sexual identity of conspecifics through volatile chemical signals in a territorial salamander. *Ethology*, 113, 214–222.

Dantzer, B. J., & Jaeger, R. G. (2007b). Male red-backed salamanders can determine the reproductive status of conspecific females through volatile chemical signals. *Herpetologica*, 63, 176–183.

Darwin, C. (1872). *The Expression of the Emotions in Man and Animals*. London: Murray.

David, R. S., & Jaeger, R. G. (1981). Prey location through chemical cues by a terrestrial salamander. *Copeia*, 1981, 435–440.

Davies, N. B. (1978). Ecological questions about territorial behaviour. In J. R. Krebs & N. B. Davies (Eds.), *Behavioural Ecology: An Evolutionary Approach*, 1st ed. (pp. 312–350). Sunderland, MA: Sinauer Associates.

Davies, N. B., & Houston, A. I. (1984). Territory economics. In J. R. Krebs & N. B. Davies (Eds.), *Behavioural Ecology: An Evolutionary Approach*, 2nd ed. (pp. 148–169). Sunderland, MA: Sinauer Associates.

Dawkins, M. S., & Guilford, T. (1991). The corruption of honest signalling. *Animal Behaviour*, 41, 865–873.

Dawkins, R., & Krebs, J. R. (1979). Arms races between and within species. *Proceedings of the Royal Society B: Biological Sciences*, 205, 489–511.

Deitloff, J., Alcorn, M. A., & Graham, S. P. (2014). Variation in mating systems of salamanders: Mate guarding or territoriality? *Behavioural Processes*, 106, 111–117.

Del Balso, H. L. (1995). Predation, signal honesty and territoriality by the salamander *Desmognathus fuscus*. MS thesis, University of Louisiana at Lafayette.

de Waal, F. B. M., & Aureli, F. (1997). Conflict resolution and distress alleviation in monkeys and apes. In C. S. Carter, I. I. Lenderhendler, & B. Kirkpatrick (Eds.), *The Integrative Neurobiology of Affiliation* (pp. 317–328). Albany: New York Academy of Sciences.

Duhaime-Ross, A., Martel, G., & Laberge, F. (2013). Sensory determinants of agonistic interactions in the red-backed salamander, *Plethodon cinereus*. *Behaviour*, 150, 1467–1489.

Dukas, R. (1998). *Cognitive Ecology: The Evolutionary Ecology of Information Processing and Decision Making*. Chicago: University of Chicago Press.

Eberhard, W. G. (1996). *Female Control: Sexual Selection by Cryptic Female Choice*. Princeton, NJ: Princeton University Press.

Emlen, J. M. (1966). Role of time and energy in food preferences. *American Naturalist*, 100, 611–617.

Estabrook, G. F., & Dunham, A. E. (1976). Optimal diet as a function of absolute abundance, relative abundance, and relative value of available prey. *American Naturalist*, 110, 401–413.

Faragher, S. G., & Jaeger, R. G. (1997). Distributions of adult and juvenile red-backed salamanders: Testing new hypotheses regarding territoriality. *Copeia*, 1997, 410–414.

Feigenson, L., Carey, S., & Hauser, M. D. (2002). The representations underlying infants' choice of more: Object files vs. analog magnitudes. *Psychological Science*, 13, 150–156.

Figura, A. 2007. Behavioral interactions between red-backed salamanders (*Plethodon cinereus*) and amaurobiid spiders (genus *Wadotes*). MS thesis, John Carroll University, University Heights, OH.

Fraser, D. F. (1976). Empirical evaluation of the hypothesis of food competition in salamanders of the genus *Plethodon*. *Ecology*, 57, 459–471.

Gabor, C. R. (1995). Correlational test of Mathis' hypothesis that bigger salamanders have better territories. *Copeia*, 1995, 729–735.

Gabor, C. R., & Jaeger, R. G. (1995). Resource quality affects the agonistic behaviour of territorial salamanders. *Animal Behaviour*, 49, 71–79.

Gabor, C. R., & Jaeger, R. G. (1999). When salamanders misrepresent threat signals. *Copeia*, 1999, 1123–1126.

Gall, S. B., Anthony, C. D., & Wicknick, J. A. (2003). Do behavioral interactions between salamanders and beetles indicate a guild relationship? *American Midland Naturalist*, 149, 363–374.

Gergits, W. F., & Jaeger, R. G. (1990a). Field observations of the behavior of the red-backed salamander (*Plethodon cinereus*): Courtship and agonistic interactions. *Journal of Herpetology*, 24, 93–95.

Gergits, W. F., & Jaeger, R. G. (1990b). Site attachment by the red-backed salamander, *Plethodon cinereus*. *Journal of Herpetology*, 24, 91–93.

Gibbons, M. E., Ferguson, A. M., & Lee, D. R. (2005). Both learning and heritability affect foraging behaviour of red-backed salamanders, *Plethodon cinereus*. *Animal Behaviour*, 69, 721–732.

Gibbons, M. E., Ferguson, A. M., Lee, D. R., & Jaeger, R. G. (2003). Mother–offspring discrimination in the red-backed salamander may be context dependent. *Herpetologica*, 59, 322–333.

Gifford, M. E., & Kozak, K. H. (2012). Islands in the sky or squeezed at the top? Ecological causes of elevational range limits in montane salamanders. *Ecography*, 35, 193–203.

Gillette, J. R. (2003). Population ecology, social behavior, and intersexual differences in a natural population of red-backed salamanders: A long-term field study. PhD diss., University of Louisiana at Lafayette.

Gillette, J. R., Jaeger, R. G., & Peterson, M. G. (2000). Social monogamy in a territorial salamander. *Animal Behaviour*, 59, 1241–1250.

Gillette, J. R., Kolb, S. E., Smith, J. A., & Jaeger, R. G. (2000). Pheromonal attractions to particular males in female redback salamanders (*Plethodon cinereus*). In R. C. Bruce, R. G. Jaeger, & L. D. Houck (Eds.), *The Biology of Plethodontid Salamanders* (pp. 431–440). New York: Kluwer Academic/Plenum.

Gillette, J. R., & Peterson, M. G. (2001). The benefits of transparency: Candling as a simple method for determining sex in red-backed salamanders (*Plethodon cinereus*). *Herpetological Review*, 32, 233–235.

Golabek, K. A., Ridley, A. R., & Radford, A. N. (2012). Food availability affects strength of seasonal territorial behaviour in a cooperatively breeding bird. *Animal Behaviour*, 83, 613–619.

Gollmann, G., Gollmann, B., & Jaeger, R. G. (2004). Spatial distribution of red-backed salamanders *Plethodon cinereus* (Green, 1818) in relation to microhabitat structure. *Herpetozoa*, 16, 169–171.

Gosling, L. M. (1990). Scent marking by resource holders: Alternative mechanisms for advertising the cost of competition. In D. W. McDonald, D. Müller-Schwarze, & S. E. Natynczuk (Eds.), *Chemical Signals in Vertebrates V* (pp. 315–328). Oxford: Oxford University Press.

Griffis, M. R., & Jaeger, R. G. (1998). Competition leads to an extinction-prone species of salamander: Interspecific territoriality in a metapopulation. *Ecology*, 79, 2494–2502.

Guffey. C., MaKinster, J. G., & Jaeger, R. G. (1998). Familiarity affects interactions between potentially courting territorial salamanders. *Copeia*, 1998, 205–208.

Hailman, J. P., & Jaeger, R. G. (1974). Phototactic responses to spectrally dominant stimuli and use of colour vision by adult anuran amphibians: A comparative survey. *Animal Behaviour*, 22, 757–795.

Hardouin, L. A., Tabel, P., & Bretagnolle, V. (2006). Neighbour–stranger discrimination in the little owl, *Athene noctua*. *Animal Behaviour*, 72, 105–112.

Hass, C. A. (1985). Geographic protein variation in the red-backed salamander (*Plethodon cinereus* Green) from the southern part of its range. MS thesis, University of Maryland, College Park, MD.

Hauser, M. D., Carey, S., & Hauser, L. (2000). Spontaneous number representation in semi-free-ranging rhesus monkeys. *Proceedings of the Royal Society B: Biological Sciences*, 267, 829–833.

Heisenberg, W. (1958). *Physics and Philosophy*. New York: Harper.

Hickerson, C.-A. M., & Anthony, C. D., & Walton, B. M. (2012). Interactions among forest-floor guild members in structurally simple microhabitats. *American Midland Naturalist*, 168, 30–42.

Hickerson, C.-A. M., Anthony, C. D., & Wicknick, J. A. (2004). Behavioral interactions between salamanders and centipedes: Competition in divergent taxa. *Behavioral Ecology*, 15, 679–686.
Highton, R. (1962). Revision of North American salamanders of the genus *Plethodon*. *Bulletin of the Florida State Museum: Biological Sciences*, 6, 235–367.
Highton, R. (1971). Distributional interactions among eastern North American salamanders of the genus *Plethodon*. In P. C. Holt (Ed.), *The Distributional History of the Biota of the Southern Appalachians* (pp. 139–188). Virginia Polytechnic Institute and State University Research Division Monograph 4. Blacksburg: Virginia Polytechnic Institute.
Highton, R., & Larson, A. (1979). The genetic relationships of the salamanders of the genus *Plethodon*. *Systematic Zoology*, 28, 579–599.
Highton, R., & Worthington, R. D. (1967). A new salamander of the genus *Plethodon* from Virginia. *Copeia*, 1967, 617–626.
Hill, J., Formanowicz, D. R., & Jaeger, R. G. (1982). Patch foraging by a terrestrial salamander. *Journal of Herpetology*, 16, 405–408.
Hom, C. L., Jaeger, R. G., & Willits, N. H. (1997). Courtship behaviour in male red-backed salamanders: The ESS dating game. *Animal Behaviour*, 54, 715–724.
Horne, E. A. (1988). Aggressive behavior of female red-backed salamanders. *Herpetologica*, 44, 203–209.
Horne, E. A., & Jaeger, R. G. (1988). Territorial pheromones of female red-backed salamanders. *Ethology*, 78, 143–152.
Hume, D. (1995). *An Inquiry Concerning Human Understanding*. New York: Liberal Arts Press. (Originally published 1748 as *Philosophical Essays Concerning Human Understanding*).
Isaacson, W. (2007). *Einstein: His Life and Universe*. New York: Simon & Schuster.
Ivanov, K., Lockhart, O. M., Keiper, J., & Walton, B. M. (2011). Status of the exotic ant *Nylanderia flavipes* (Hymenoptera: Formicidae) in northeastern Ohio. *Biological Invasions*, 13, 1945–1950.
Jaeger, R. G. (1970). Potential extinction through competition between two species of terrestrial salamanders. *Evolution*, 24, 632–642.
Jaeger, R. G. (1971a). Competitive exclusion as a factor influencing the distributions of two species of terrestrial salamanders. *Ecology*, 52, 632–637.
Jaeger, R. G. (1971b). Moisture as a factor influencing the distributions of two species of terrestrial salamanders. *Oecologia*, 6, 191–207.
Jaeger, R. G. (1972). Food as a limited resource in competition between two species of terrestrial salamanders. *Ecology*, 53, 535–546.
Jaeger, R. G. (1974a). Competitive exclusion: Comments on survival and extinction of species. *BioScience*, 24, 33–39.
Jaeger, R. G. (1974b). Interference or exploitation? A second look at competition between salamanders. *Journal of Herpetology*, 8, 191–194.
Jaeger, R. G. (1978). Plant climbing by salamanders: Periodic availability of plant-dwelling prey. *Copeia*, 1978, 686–691.
Jaeger, R. G. (1979). Seasonal spatial distributions of the terrestrial salamander *Plethodon cinereus*. *Herpetologica*, 35, 90–93.
Jaeger, R. G. (1980a). Density-dependent and density-independent causes of extinction of a salamander population. *Evolution*, 34, 617–621.

Jaeger, R. G. (1980b). Fluctuations in prey availability and food limitation for a terrestrial salamander. *Oecologia*, 44, 335–341.

Jaeger, R. G. (1980c). Microhabitats of a terrestrial forest salamander. *Copeia*, 1980, 265–268.

Jaeger, R. G. (1981a). Birds as inefficient predators on terrestrial salamanders. *American Naturalist*, 117, 835–837.

Jaeger, R. G. (1981b). Dear enemy recognition and the costs of aggression between salamanders. *American Naturalist*, 117, 962–974.

Jaeger, R. G. (1981c). Diet diversity and clutch size of aquatic and terrestrial salamanders. *Oecologia*, 48, 190–193.

Jaeger, R. G. (1984). Agonistic behavior of the red-backed salamander. *Copeia*, 1984, 309–314.

Jaeger, R. G. (1986). Pheromonal markers as territorial advertisement by terrestrial salamanders. In D. Duvall, D. Müller-Schwarze, & R. M. Silverstein (Eds.), *Chemical Signals in Vertebrates 4* (pp. 191–203). New York: Plenum.

Jaeger, R. G. (1990). Territorial salamanders evaluate size and chitinous content of arthropod prey. In R. N. Hughes (Ed.), *Behavioural Mechanisms of Food Selection* (pp. 111–126). NATO ASI Series, Subseries G: Ecological Sciences. Heidelberg, Germany: Springer-Verlag.

Jaeger, R. G. (1992). Housing, handling, and nutrition of salamanders. In D. O. Schaeffer, K. M. Kleinow, & L. Krulish (Eds.), *The Care and Use of Amphibians, Reptiles and Fish in Research* (pp. 25–29). Baltimore, MD: Scientists Center for Animal Welfare.

Jaeger, R. G. (1994a). Patch sampling. In W. R. Heyer, M. A. Donnelly, R. W. McDiarmid, L. A. Hayek, & M. S. Foster (Eds.), *Measuring and Monitoring Biological Diversity: Standard Methods for Amphibians* (pp. 107–109). Biological Diversity Handbook Series. Washington, DC: Smithsonian Institution Press.

Jaeger, R. G. (1994b). Transect sampling. In W. R. Heyer, M. A. Donnelly, R. W. McDiarmid, L. A. Hayek, & M. S. Foster (Eds.), *Measuring and Monitoring Biological Diversity: Standard Methods for Amphibians* (pp. 103–107). Biological Diversity Handbook Series. Washington, DC: Smithsonian Institution Press.

Jaeger, R. G., & Barnard, D. E. (1981). Foraging tactics of a terrestrial salamander: Choice of diet in structurally simple environments. *American Naturalist*, 117, 639–664.

Jaeger, R. G., Barnard, D. E., & Joseph, R. G. (1982). Foraging tactics of a terrestrial salamander: Assessing prey density. *American Naturalist*, 119, 885–890.

Jaeger, R. G., & Forester, D. C. (1993). Social behavior of plethodontid salamanders. *Herpetologica*, 49, 163–175.

Jaeger, R. G., Fortune, D., Hill, G., Palen, A., & Risher, G. (1993). Salamander homing behavior and territorial pheromones: Alternative hypotheses. *Journal of Herpetology*, 27, 236–239.

Jaeger, R. G., & Gabor, C. R. (1993). Intraspecific chemical communication by a territorial salamander via the postcloacal gland. *Copeia*, 1993, 1171–1174.

Jaeger, R. G., Gabor, C. R., & Wilbur, H. M. (1998). An assemblage of salamanders in the southern Appalachian Mountains: Competitive and predatory behavior. *Behaviour*, 135, 795–821.

Jaeger, R. G., & Gergits, W. F. (1979). Intra- and interspecific communication in salamanders through chemical signals on the substrate. *Animal Behaviour*, 27, 150–156.

Jaeger, R. G., Gillette, J. R., & Cooper, R. C. (2002). Sexual coercion in a territorial salamander: males punish socially polyandrous female partners. *Animal Behaviour*, 63, 871–877.
Jaeger, R. G., Goy, J. M., Tarver, M., & Márquez, C. E. (1986). Salamander territoriality: Pheromonal markers as advertisement by males. *Animal Behaviour*, 34, 860–864.
Jaeger, R. G., & Hailman, J. P. (1971). Two types of phototactic behavior in anuran amphibians. *Nature*, 230, 189–190.
Jaeger, R. G., & Halliday, T. R. (1998). On confirmatory versus exploratory research. *Herpetologica*, 54, S64–S66.
Jaeger, R. G., & Inger, R. F. (1994). Quadrat sampling. In W. R. Heyer, M. A. Donnelly, R. W. McDiarmid, L. A. Hayek, & M. S. Foster (Eds.), *Measuring and Monitoring Biological Diversity: Standard Methods for Amphibians* (pp. 97–102). Biological Diversity Handbook Series. Washington, DC: Smithsonian Institution Press.
Jaeger, R. G., Joseph, R. G., & Barnard, D. E. (1981). Foraging tactics of a terrestrial salamander: Sustained yield in territories. *Animal Behaviour*, 29, 1100–1105.
Jaeger, R. G., Kalvarsky, D., & Shimizu, N. (1982). Territorial behaviour of the red-backed salamander: Expulsion of intruders. *Animal Behaviour*, 30, 490–496.
Jaeger, R. G., & Lucas, J. (1990). On evaluation of foraging strategies through estimates of reproductive success. In R. N. Hughes (Ed.), *Behavioural Mechanisms of Food Selection*. (pp. 83–94). NATO ASI Series, Subseries G: Ecological Sciences. Heidelberg, Germany: Springer-Verlag.
Jaeger, R. G., Nishikawa, K. C. B., & Barnard, D. E. (1983). Foraging tactics of a terrestrial salamander: Costs of territorial defence. *Animal Behaviour*, 31, 191–198.
Jaeger, R. G., & Peterson, M. G. (2002). Familiarity affects agonistic interactions between female red-backed salamanders. *Copeia*, 2002, 865–869.
Jaeger, R. G., Peterson, M. G., & Gillette, J. R. (2000). A model of alternative mating strategies in the redback salamander, *Plethodon cinereus*. In R. C. Bruce, R. G. Jaeger, & L. D. Houck (Eds.), *The Biology of Plethodontid Salamanders* (pp. 441–450). New York: Kluwer Academic/Plenum.
Jaeger, R. G., Peterson, M. G., Gollmann, G., Gollmann, B., & Townsend, V. R. Jr. (2001). Salamander social strategies: Living together in female–male pairs. *Journal of Herpetology*, 35, 335–338.
Jaeger, R. G., Prosen, E. D., & Adams, D. C. (2002). Character displacement and aggression in two species of terrestrial salamanders. *Copeia*, 2002, 391–401.
Jaeger, R. G., & Rubin, A. M. (1982). Foraging tactics of a terrestrial salamander: Judging prey profitability. *Journal of Animal Ecology*, 51, 167–176.
Jaeger, R. G., & Schwarz, J. K. (1991). Gradational threat postures by the red-backed salamander. *Journal of Herpetology*, 25, 112–114.
Jaeger, R. G., Schwarz, J., & Wise, S. E. (1995). Territorial male salamanders have foraging tactics attractive to gravid females. *Animal Behaviour*, 49, 633–639.
Jaeger, R. G., Wang, H. G., Kohn, N. R., & Cross, H. S. (2005). Displacement of territorial aggression by male red-backed salamanders. *Journal of Herpetology*, 39, 653–656.
Jaeger, R. G., Wicknick, J. A., Griffis, M. R., & Anthony, C. D. (1995). Socioecology of a terrestrial salamander: Juveniles enter adult territories during stressful foraging periods. *Ecology*, 76, 533–543.
Jaeger, R. G., & Wise, S. E. (1991). A reexamination of the male salamander "sexy faeces hypothesis." *Journal of Herpetology*, 25, 370–373.

Joseph, L. L., Cope, C. T., Kohn, N. R., & Jaeger, R. G. (2005). The influence of familiarity on socially monogamous behavior of red-backed salamanders. *Herpetologica*, 61, 343–348.

Kaplan, D. L. (1977). Exploitative competition in salamanders: Test of a hypothesis. *Copeia*, 1977, 234–238.

Karuzas, J. M., Maerz, J. C., & Madison, D. M. (2004). An alternative hypothesis for the primary function of a proposed mate assessment behaviour in red-backed salamanders. *Animal Behaviour*, 68, 489–494.

Kawecki, T. J. (2010). Evolutionary ecology of learning: Insights from fruit flies. *Population Ecology*, 52, 15–25.

Kleeberger, S. R., & Werner, J. K. (1982). Home range and homing behavior of *Plethodon cinereus* in northern Michigan. *Copeia*, 1982, 409–415.

Kohn, N. R. (2008). Conspecific "distractors" do not affect individual recognition memory in territorial male red-backed salamanders. In N. R. Kohn, Tests of individual recognition memory in male red-backed salamanders (*Plethodon cinereus*) (pp. 80–105). PhD diss., University of Louisiana at Lafayette, Lafayette, Louisiana.

Kohn, N. R., Deitloff, J. M., Dartez, S. F., Wilcox, M. M., & Jaeger, R. G. (2013). Memory of conspecifics in male salamanders *Plethodon cinereus*: Implications for territorial defense. *Current Zoology*, 59, 326–334.

Kohn, N. R., & Jaeger, R. G. (2009). Male salamanders remember individuals based on chemical or visual cues. *Behaviour*, 146, 1485–1498.

Kohn, N. R., Jaeger, R. G., & Franchebois, J. (2005). Effects of intruder number and sex on territorial behavior of female red-backed salamanders (*Plethodon cinereus*: Plethodontidae). *Journal of Herpetology*, 39, 645–648.

Koster, E. P., Degel, J., & Piper, D. (2002). Proactive and retroactive interference in implicit odor memory. *Chemical Senses*, 27, 191–206.

Kraemer, A. C., & Adams, D. C. (2014). Predator perception of Batesian mimicry and conspicuousness in a salamander. *Evolution*, 68, 1197–1206.

Kraemer, A. C., Serb, J. M., & Adams, D. C. (2015). Batesian mimics influence the evolution of conspicuousness in an aposematic salamander. *Journal of Evolutionary Biology*, 28, 1016–1023.

Krebs, J. R. (1978). Optimal foraging: Decision rules for predators. In J. R. Krebs & N. B. Davies (Eds.), *Behavioural Ecology: An Evolutionary Approach* (pp. 23–63). Sunderland, MA: Sinauer Associates.

Krebs, J. R. (1982). Territorial defence in the great tit (*Parus major*): Do residents always win? *Behavioral Ecology and Sociobiology*, 11, 185–194.

Krusche, P., Uller, C., & Dicke, U. (2010). Quantity discrimination in salamanders. *Journal of Experimental Biology*, 213, 1822–1828.

Lakatos, I. (1970). Falsification and the methodology of scientific research programmes. In I. Lakatos & A. Musgrave (Eds.), *Criticism and the Growth of Knowledge* (pp. 189–195). Cambridge, UK: Cambridge University Press.

Lancaster, D. L. (1994). The role of substrate odors in the predator–prey relationship of *Diadophis punctatus* and *Plethodon cinereus*: Tests of the detection of potential predators and the detection of potential prey. PhD diss., University of Louisiana at Lafayette.

Lancaster, D., & Jaeger, R. (1995). Rules of engagement for adult salamanders in territorial conflicts with heterospecific juveniles. *Behavioral Ecology and Sociobiology*, 37, 25–29.

Lancaster, D. L., & Wise, S. E. (1996). Differential response by the ringneck snake, *Diadophis punctatus*, to odors of tail-autotomizing prey. *Herpetologica*, 52, 98–108.
Lang, C., & Jaeger, R. G. (2000). Defense of territories by male–female pairs in the red-backed salamander (*Plethodon cinereus*). *Copeia*, 2000, 169–177.
Liebgold, E. B. (2014). The influence of social environment: Behavior of unrelated adults affects future juvenile behaviors. *Ethology*, 120, 388–399.
Liebgold, E. B., & Cabe, P. R. (2008). Familiarity with adults, but not relatedness, affects the growth of juvenile red-backed salamanders (*Plethodon cinereus*). *Behavioral Ecology and Sociobiology*, 63, 277–284.
Liebgold, E. B., Cabe, P. R., Jaeger, R. G., & Leberg, P. L. (2006). Multiple paternity in a salamander with socially monogamous behaviour. *Molecular Ecology*, 15, 4153–4160.
Liebgold, E. B., & Jaeger, R. G. (2007). Juvenile movements and potential inter-age class associations of red-backed salamanders. *Herpetologica*, 63, 51–55.
Lucon-Xiccato, T., Miletto Petrazzini, M. E., Agrillo, C., & Bisazza, A. (2015). Guppies discriminate between two quantities of food items but prioritize item size over total amount. *Animal Behaviour*, 107, 183–191.
MacArthur, R. H. (1958). Population ecology of some warblers of northeastern coniferous forests. *Ecology*, 39, 599–619.
MacArthur, R. H. (1970). Species packing and competitive equilibrium for many species. *Theoretical Population Biology*, 1, 1–11.
MacArthur, R. H., & Levins, R. (1964). Competition, habitat selection, and character displacement in a patchy environment. *Proceedings of the National Academy of Sciences USA*, 51, 1207–1210.
MacArthur, R. H., & Pianka, E. R. (1966). On optimal use of a patchy environment. *American Naturalist*, 100, 603–609.
MacArthur, R. H., & Wilson, E. O. (1967). *The Theory of Island Biogeography*. Princeton, NJ: Princeton University Press.
Madison, D. M. (1969). Homing behaviour of the red-checked salamander, *Plethodon jordani*. *Animal Behaviour*, 17, 25–39.
Madison, D. M. (1975). Intraspecific odor preferences between salamanders of the same sex: dependence on season and proximity of resident. *Canadian Journal of Zoology*, 53, 1356–1361.
Madison, D. M., Maerz, J. C., & McDarby, J. H. (1999). Optimization of predator avoidance by salamanders using chemical cues: Diet and diel effects. *Ethology*, 105, 1073–1086.
Madison, D., Sullivan, A., Maerz, J., McDarby, J., & Rohr, J. (2002). A complex, cross-taxon, chemical releaser of antipredator behavior in amphibians. *Journal of Chemical Ecology*, 28, 2271–2282.
Maerz, J. C., & Madison, D. M. (2000). Environmental variation and territorial behavior in a terrestrial salamander. In R. C. Bruce, R. G. Jaeger, & L. D. Houck (Eds.), *The Biology of Plethodontid Salamanders* (pp. 395–406). New York: Kluwer Academic/Plenum.
Maerz, J. C., Karuzas, J. M., Madison, D. M., & Blossey, B. (2005). Introduced invertebrates are important prey for a generalist predator. *Diversity and Distributions*, 11, 83–90.
Maksimowich, D. S., & Mathis, A. (2000). Parasitized salamanders are inferior competitors for territories and food resources. *Ethology*, 106, 319–329.

Martin, S. B., Jaeger, R. G., & Prosen, E. D. (2005). Territorial red-backed salamanders can detect volatile pheromones from intruders. *Herpetologica*, 61, 29–35.

Mathis, A. (1989). Do seasonal spatial distributions in a terrestrial salamander reflect reproductive behavior or territoriality? *Copeia*, 1989, 788–791.

Mathis, A. (1990a). Territoriality in a terrestrial salamander: The influence of resource quality and body size. *Behaviour*, 112, 162–174.

Mathis, A., (1990b). Territorial salamanders assess sexual and competitive information using chemical signals. *Animal Behaviour*, 40, 953–962.

Mathis, A., (1991a). Large male advantage for access to females: Evidence of male–male competition and female discrimination in a territorial salamander. *Behavioral Ecology and Sociobiology*, 29, 133–138.

Mathis, A., (1991b). Territories of male and female terrestrial salamanders: Costs, benefits, and intersexual spatial associations. *Oecologia*, 86, 433–440.

Mathis A., Jaeger, R. G., Keen, W. H., Ducey, P. K., Walls, S. C., & Buchanan, B. C. (1995). Aggression and territoriality by salamanders and a comparison with the territorial behavior of frogs. In H. Heatwole & B. K. Sullivan (Eds.), *Amphibian Biology:* Vol. 2, *Social Behaviour* (pp. 633–676). Chipping Norton, Australia: Surrey Beatty & Sons.

Mathis, A., & Simons, R. R. (1994). Size-dependent responses of resident male red-backed salamanders to chemical stimuli from conspecifics. *Herpetologica*, 50, 335–344.

Maynard Smith, J. (1982). *Evolution and the Theory of Games*. Cambridge, UK: Cambridge University Press.

Maynard Smith, J., & Parker, G. A. (1976). The logic of asymmetric contests. *Animal Behaviour*, 24, 159–175.

McGavin, M. (1978). Recognition of conspecific odors by the salamander *Plethodon cinereus*. *Copeia*, 1978, 356–358.

Meche, G., & Jaeger, R. G. (2002). Associations of male red-backed salamanders with tail-intact versus tail autotomized females during the courtship season. *Journal of Herpetology*, 36, 532–535.

Merchant, H. C. (1970). Estimated energy budget of the red-backed salamander, *Plethodon cinereus*. PhD diss., Rutgers University, New Brunswick, NJ.

Moore, A. L., Williams, C. E., Martin, T. H., & Moriarity, W. J. (2001). Influence of season, geomorphic surface and cover item on capture, size and weight of *Desmognathus ochrophaeus* and *Plethodon cinereus* in Allegheny Plateau riparian forests. *American Midland Naturalist*, 145, 39–45.

Myers, E. M., & Adams, D. C. (2008). Morphology is decoupled from interspecific competition in *Plethodon* salamanders in the Shenandoah Mountains, USA. *Herpetologica*, 64, 281–289.

Ng, M. Y., & Wilbur, H. M. (1995). The cost of brooding in *Plethodon cinereus*. *Herpetologica*, 51, 1–8.

Nunes, V. da S. (1988). Feeding asymmetry affects territorial disputes between males of *Plethodon cinereus*. *Herpetologica*, 44, 386–391.

Nunes, V. da S., & Jaeger, R. G. (1989). Salamander aggressiveness increases with length of territorial ownership. *Copeia*, 1989, 712–718.

Page, R. B., & Jaeger, R. G. (2004). Multimodal signals, imperfect information, and identification of sex in red-backed salamanders (*Plethodon cinereus*). *Behavioral Ecology and Sociobiology*, 56, 132–139.

Palphramand, K. L., & White, P. C. L. (2007). Badgers, *Meles meles*, discriminate between neighbour, alien, and self scent. *Animal Behaviour*, 74, 429–436.
Paluh, D. J., Eddy, C., Ivanov, K., Hickerson, C. M., & Anthony, C. D. (2015). Selective foraging on ants by a terrestrial polymorphic salamander. *American Midland Naturalist*, 174, 265–277.
Peterson, M. G. (2000). Nest, but not egg, fidelity in a territorial salamander. *Ethology*, 106, 781–794.
Peterson, M. G., Gillette, J. R., Franks, R., & Jaeger, R. G. (2000). Alternative life styles in a terrestrial salamander: do females preferentially associate with each other? In R. C. Bruce, R. G. Jaeger, & L. D. Houck (Eds.), *The Biology of Plethodontid Salamanders* (pp. 417–429). New York: Kluwer Academic/Plenum.
Petranka, J. W. (1998). *Salamanders of the United States and Canada*. Washington, DC: Smithsonian Institution Press.
Platt, J. R. (1964). Strong inference. *Science*, 146, 347–353.
Popper, K. R. (1959). *The Logic of Scientific Discovery*. New York: Basic Books.
Prosen, E. D. (2004). Sexual coercion and relationship value in the red-backed salamander (*Plethodon cinereus*). PhD diss., University of Louisiana at Lafayette.
Prosen, E. D., Jaeger, R. G., & Hucko, J. A. (2006). Sexual coercion in the salamander *Plethodon cinereus*. Is it merely a result of familiarity? *Herpetologica*, 62, 10–18.
Prosen, E. D., Jaeger, R. G., & Lee, D. R. (2004). Sexual coercion in a territorial salamander: Females punish socially polygynous male partners. *Animal Behaviour*, 67, 85–92.
Pulliam, H. R. (1988). Sources, sinks, and population regulation. *American Naturalist*, 132, 652–661.
Pyke, G. H., Pulliam, H. R., & Charnov, E. L. (1977). Optimal foraging: A selective review of theory and tests. *Quarterly Review of Biology*, 52, 137–154.
Quinn, V. S., & Graves, B. M. (1999). Space use in response to conspecifics by the red-backed salamander (*Plethodon cinereus*, Plethodontidae, Caudata). *Ethology*, 105, 993–1002.
Ransom, T. S., & Jaeger, R. G. (2006). An assemblage of salamanders in the southern Appalachian Mountains revisited: Competitive and predatory behavior? *Behaviour*, 143, 1357–1382.
Ransom, T. S. & Jaeger, R. G. (2008). Intergeneric salamander interactions across an ecotone. *Herpetologica*, 64, 20–31.
Real, L. A. (1993). Toward a cognitive ecology. *Trends in Ecology and Evolution*, 8, 413–417.
Reiter, M. K., Anthony, C. D., & Hickerson, C.-A. M. (2014). Territorial behavior and ecological divergence in a polymorphic salamander. *Copeia*, 2014, 481–488.
Rollinson, N., & Hackett, D. (2015). Experimental evaluation of agonistic behaviour, chemical communication, spacing, and intersexual associations of the red-backed salamander (*Plethodon cinereus*) near its northern range limit. *Canadian Journal of Zoology*, 93, 773–781.
Rowe, C. (1999). Receiver psychology and the evolution of multicomponent signals. *Animal Behaviour*, 58, 921–931.
Ruby, N. P. (2004). Factors influencing numerical discrimination in the red-backed salamander (*Plethodon cinereus*). MS thesis, University of Louisiana at Lafayette.
Schubert, S. N., Houck, L. D., Feldhoff, P. W., Feldhoff, R. C., & Woodley, S. K. (2008). The effects of sex on chemosensory communication in a terrestrial salamander (*Plethodon shermani*). *Hormones and Behavior*, 54, 270–277.

Seaman, J. W., & Jaeger, R. G. (1990). Statisticae dogmaticae: A critical essay on statistical practice in ecology. *Herpetologica*, 46, 337–346.

Sever, D. M. (1976). Morphology of the mental hedonic gland clusters of plethodontid salamanders (Amphibia, Urodela, Plethodontidae). *Journal of Herpetology*, 10, 227–239.

Sever, D. M. (1978a). Female cloacal anatomy of *Plethodon cinereus* and *Plethodon dorsalis* (Amphibia, Urodela, Plethodontidae). *Journal of Herpetology*, 12, 397–406.

Sever, D. M. (1978b). Male cloacal glands of *Plethodon cinereus* and *Plethodon dorsalis* (Amphibia: Plethodontidae). *Herpetologica*, 34, 1–20.

Sever, D. M. (1989). Caudal hedonic glands in salamanders of the *Eurycea bislineata* complex (Amphibia: Plethodontidae). *Herpetologica*, 45, 322–329.

Siegel, S., & Castellan, N. J. Jr. (1988). *Nonparametric Statistics for the Behavioral Sciences*. New York: McGraw-Hill.

Sih, A., Bell, A., & Johnson, J. C. (2004). Behavioral syndromes: An ecological and evolutionary overview. *Trends in Ecology & Evolution*, 19, 372–378.

Simon, G. S., & Madison, D. M., (1984). Individual recognition in salamanders: Cloacal odours. *Animal Behaviour*, 32, 1017–1020.

Simons, R. R., & Felgenhauer, B. E. (1992). Identifying areas of chemical signal production in the red-backed salamander, *Plethodon cinereus*. *Copeia*, 1992, 766–781.

Simons, R. R., Felgenhauer, B. E., & Jaeger, R. G. (1993). A technique to explore the effects of integumental glands on the behavior of salamanders. *Journal of Herpetology*, 27, 99–100.

Simons, R. R., Felgenhauer, B. E., & Jaeger, R. G. (1994). Salamander scent marks: Site of production and their role in territorial defence. *Animal Behaviour*, 48, 97–103.

Simons, R. R., Jaeger, R. G., & Felgenhauer, B. E. (1995). Juvenile territorial salamanders have active postcloacal glands. *Copeia*, 1995, 481–483.

Simons, R. R., Jaeger, R. G., & Felgenhauer, B. E. (1997). Competitor assessment and area defense by territorial salamanders. *Copeia*, 1997, 70–76.

Smith, J. N. M., & Dawkins, R. (1971). The hunting behaviour of individual great tits in relation to spatial variation in their food density. *Animal Behaviour*, 19, 695–706.

Stancher, G., Rugani, R., Regolin, L., & Vallortigara, G. (2015). Numerical discrimination by frogs (*Bombina orientalis*). *Animal Cognition*, 18, 219–229.

Stephens, D. W., & Krebs, J. R. (1986). *Foraging Theory*. Princeton, NJ: Princeton University Press.

Sullivan, A. M., Madison, D. M., & Rohr, J. R. (2003). Behavioural responses by red-backed salamanders to conspecific and heterospecific cues. *Behaviour*, 140, 553–564.

Taub, F. B. (1961). The distribution of the red-backed salamander, *Plethodon cinereus*, within the soil. *Ecology*, 42, 681–698.

Thomas, J. S., Jaeger, R. G., & Horne, E. A. (1989). Are all females welcome? Agonistic behavior of male red-backed salamanders. *Copeia*, 1989, 915–920.

Thurow, G. (1976). Aggression and competition of eastern *Plethodon* (Amphibia, Urodela, Plethodontidae). *Journal of Herpetology*, 10, 277–291.

Toll, S. J., Jaeger, R. G., & Gillette, J. R. (2000). Socioecology of a terrestrial salamander: Females and males as territorial residents and invaders. *Copeia*, 2000, 276–281.

Townsend, V. R. Jr., Akin, J. A., & Jaeger, R. G. (1998). The significance of small body size in territorial defense in the red-backed salamander, *Plethodon cinereus*. *Journal of Herpetology*, 32, 579–581.

Townsend, V. R. Jr., & Jaeger, R. G. (1998). Territorial conflicts over prey: Domination by large male salamanders. *Copeia*, 1998, 725–729.
Tristram, D. A. (1977). Intraspecific olfactory communication in the terrestrial salamander *Plethodon cinereus*. *Copeia*, 1977, 597–600.
Uller, C., Jaeger, R. G., Guidry, G., & Martin, C. (2003). Salamanders (*Plethodon cinereus*) go for more: Rudiments of number in an amphibian. *Animal Cognition*, 6, 105–112.
Walls, S. C., Mathis, A., Jaeger, R. G., & Gergits, W. F. (1989). Male salamanders with high quality diets have faeces attractive to females. *Animal Behaviour*, 38, 546–548.
Wells, K. D. (2007). *The Ecology and Behavior of Amphibians*. Chicago: University of Chicago Press.
Wicknick, J. R. (1995). Interspecific competition and territoriality between a widespread species of salamander and a species with a limited range. PhD diss., University of Louisiana at Lafayette.
Wilcox, M. M. (2006). The effects of food resources on social monogamy in pairs of red-backed salamanders (*Plethodon cinereus*). MS thesis, University of Louisiana at Lafayette.
Wilson, E. O. (1975). *Sociobiology: The New Synthesis*. Cambridge, MA: Harvard University Press.
Wise, S. E. (1995). Variation in the life history and sociobiology of a territorial salamander. PhD diss., University of Louisiana at Lafayette.
Wise, S. E., & Jaeger, R. G. (1998). The influence of tail autotomy on agonistic behaviour in a territorial salamander. *Animal Behaviour*, 55, 1707–1716.
Wise, S. E., & Jaeger, R. G. (2016). Seasonal and geographic variation in territorial conflicts by male red-backed salamanders. *Behaviour*, 153, 187–207.
Wise, S. E., Verret, F. D., & Jaeger, R. G. (2004). Tail autotomy in territorial salamanders influences scent marking by residents and behavioral responses of intruders to resident chemical cues. *Copeia*, 2004, 165–172.
Wittenberger, J. F., & Tilson, R. L. (1980). The evolution of monogamy: Hypotheses and evidence. *Annual Review of Ecology and Systematics*, 11, 197–232.
Woodley, S. (2015). Chemosignals, hormones, and amphibian reproduction. *Hormones and Behavior*, 68, 3–13.
Woodley, S. K., & Lacy, E. L. (2010). An acute stressor alters steroid hormone levels and activity but not sexual behavior in male and female Ocoee salamanders (*Desmognathus ocoee*). *Hormones and Behavior*, 58, 427–432.
Wrobel, D. J., Gergits, W. F., & Jaeger, R. G. (1980). An experimental study of interference competition among terrestrial salamanders. *Ecology*, 61, 1034–1039.
Ziemba, J. L., Cameron, A. C., Peterson, K., Hickerson, C.-A. M., & Anthony, C. D. (2015). Invasive Asian earthworms of the genus *Amynthas* alter microhabitat use by terrestrial salamanders. *Canadian Journal of Zoology*, 93, 805–811.

SUBJECT INDEX

www.ingramcontent.com/pod-product-compliance
Ingram Content Group UK Ltd.
Pitfield, Milton Keynes, MK11 3LW, UK
UKHW021453280726
14060UKWH00001BA/381